BR 4007

Guide to Ship Firefighting

Supersedes BR 4007 Guide to Ship Firefighting 1992

October 1994

By Command of the Defence Council

Ministry of Defence
Commander-in-Chief Fleet
CinCFleet/FSAG/P4007/1

HMSO

First published 1988
Fourth edition 1994

ISBN 0 11 772683 4

The BR 2170 series of publications mentioned within this book is not available outside the MOD.

SPONSOR

This publication is sponsored by the **Commander in Chief Fleet**. All correspondence concerning this publication is to be forwarded to the **Operational Publications Authority**:

CINCFLEET/PFSA
Fleet Staff Authors Group
Pepys Building
HMS COLLINGWOOD
FAREHAM
Hants
PO14 1AS

Copies to the **Lead Authority**:

Commander NBCD
PHOENIX NBCD School
HMS EXCELLENT
Portsmouth
Hants
PO2 8ER

Foreword

This edition supersedes the 3rd edition 1992 (Green cover) which should be destroyed.

Guide to Ship Firefighting, (BR 4007) is available in all new entry training establishments to Officers and Ratings during their Pre-joining Training (PJT) Firefighting Course.

In addition copies will be held by HM Ships, Submarines and RFA's. These will be distributed to Messes and also held by Divisional Officers in order to ensure that all prospective readers will find a copy readily available.

The aim of this book is to provide, in and easily understood format, the knowledge of fire prevention and firefighting which is taught during a PJT Firefighting course.

Everyone in the RN and RFA is a potential fire fighter and, for the safety of their shipmates and the safety of the ship everyone *must be constantly aware of the fire hazards and be prepared to deal, effectively, with an outbreak of fire.*

CONTENTS

CHAPTER 1

CAUSES OF FIRE

0101. Nature of Fire

Fuel, heat and oxygen must be in contact before a fire can be started. These can be represented by the three sides of a triangle and a fire cannot start, or continue, if one of these is absent or removed.

Fig 1-1. Fire Triangle

0102. Components of a Fire

a. **Fuel.** Fuel can be solid or liquid, which, when heated gives off flammable vapours. Fuel can also be a gas which starts to burn when its ignition temperature is reached. Examples are, paper, wood, cardboard, paint, oils, acetylene, propane gas.

b. **Heat.** Heat can be transferred by:

(1) *Radiation* from any heating appliance, flames or explosion; or

(2) *Conduction* through any suitable material such as steel or aluminium, decks or bulkheads; or

(3) *Convection* via gases, liquids or hot air circulated through ventilation trunking, lift shafts, etc.

c. **Oxygen.** A supply of this gas, which is essential for combustion to take place, is available in the air.

Causes of fire

0103. Types of Fire

a. The British Standard method of classifying fires is by the nature of the fuel involved and is as follows:

(1) **Class A** Involving solid materials usually of an organic nature.

(2) **Class B** Involving liquids or liquefiable solids.

(3) **Class C** Involving flammable gases.

(4) **Class D** Involving metals.

b. Most shipboard fires involve a variety of fuels and these classifications are not always appropriate. In order that the appropriate firefighting equipment is provided at the scene of the fire (SOF) and the appropriate drills are initiated, shipboard fires are categorised as follows:

(1) **Category A - Solid Fuel Fire.** Wood paper, cardboard fabrics and many other similar materials.

(2) **Category B - Oil Fuel Fires.** Lubricating oil, hydraulic oil, kerosene, diesel, petrol, etc.

(3) **Category C - Electrical Fire.** Electricity does not burn. However, when the origin of a fire is electrical (usually a fault in a live circuit which has generated enough heat to ignite a combustible material) it is referred to as an **electrical fire** which is a warning to firefighters. They will then take the necessary measures to locate local circuits associated with any fire. Any circuits/ equipments which are not under local control must be referred to HQ1/SCC/MSB for the necessary action. Once the necessary circuits have been isolated, the fire is fought with the techniques and equipment appropriate to the material on fire.

0104. Principles of Extinction

a. By cooling - the burning material is cooled below its ignition temperature, usually using water.

b. By smothering - by excluding oxygen (air) from the burning material usually using CO_2, steam or foam.

c. By removal of the fuel - accomplished by the removal of the combustible material or turning off the liquid fuel supply.

d. By disrupting the chemical reaction of the burning process - by using halon gas (BCF/BTM).

CHAPTER 2

FIRE PREVENTION

0201. Introduction

Fires onboard can destroy accommodation, stores, equipment, lives, and in extreme cases, the ship. Firefighting onboard is difficult and beset with dangers. When good fire prevention drills are exercised by the **whole** ship's company these difficulties and hazards are greatly reduced and can be almost eliminated. If you want to keep your **home** intact, then avoid the hazards of fire onboard by putting the **aide memoire** into practice - it is more convenient and less dangerous than fighting fires.

0202. Fire Prevention - *aide memoire*

a. All private electrical equipment onboard must be checked safe by the ME department and logged.

b. Clothes and other flammables must be kept clear of electrical radiators, hot equipment and systems.

c. Smoking should only take place in spaces, and only in situations, where it is safe to do so. **Do not smoke in bed.** Do not smoke when moving around the ship and especially when on rounds. Do not flick cigarette butts or pipe knockings over the ships side. All smoking materials must be extinguished and placed in a suitable container, eg steel gash bin.

d. Messdecks are to be kept tidy with all personal gear stowed in the kit lockers, which should be securely closed.

e. Electric irons and soldering irons are to be switched off when not in use, and are to be allowed to cool before being stowed away. These appliances **must not** be left unattended when they are switched on.

f. All paints and painting materials are to be returned to the Paint Shop/store on completion of work in accordance with Ship's Standing Orders (SSO's).

g. All other highly flammable materials, such as solvent, aerosols, some adhesives and liquids, gases and petrol, are to be used in accordance with the manufactures instructions, or current MOD(N) instructions and when not required for immediate use, to have their containers sealed. At the end of the working day, they are to be placed, and secured, in an authorized stowage in accordance with SSO's.

h. All rags, especially wet or oily ones, sawdust and wood chips, etc, are to be safely disposed of at the end of the working day in accordance with the SSO's.

i. Spillages of oil, paint, solvents, spirits or other flammable liquids are to be wiped up immediately and the dirty rags disposed of in accordance with SSO's.

j. When any compartment is vacated all unnecessary lighting and other electrical circuits are to be switched off. The door to any empty compartment should, wherever possible, be kept shut.

k. The Welding Burning and Flame Heating Control Organization controls the safety/fire prevention aspects of all such operations (see **BR 2170 Volume 1**).

l. Only safety matches are to be used onboard.

m. The use of single action cigarette lighters and the stowage of lighter fuels (petrol or butane gas) in kit lockers is not good fire prevention practice, and is to be discouraged.

n. When not in use all cleaning gear should be stowed in approved cleaning gear stowages.

o. On completion of the working day all gash bins from living spaces and waste bins from workshops must be emptied, and the waste disposed of in a safe manner. It is recommended that gash bins and spitkids in living spaces should again be emptied before 'pipe down'.

p. When in harbour, inform the OOD immediately if any quantities of oil, or other flammable liquids, are observed overside so that action can be initiated for its safe and timely removal. This is most likely to occur when your own or nearby ships are fuelling.

q. When not in use, TV sets, radios, record players, electric fans and other similar appliances are to be switched off and disconnected from the power source. Care must be taken not to obstruct ventilation of such appliances.

r. On completion of work all papers, packs and books should be stowed in drawers, lockers or other approved stowages.

s. When **No Smoking** is piped, or **No Smoking** signs are posted in an area or space - **DO NOT SMOKE!**

t. When carrying out rounds be observant and report suspicious smells; leaks from any pipes or tanks; any electrical machinery that appears to be overheating and any wrongly stowed flammables. Check for irregular stowage spaces, masts, funnel uptakes, auxiliary machinery and other similar spaces. Report any safety/firefighting equipment that is missing or incorrectly stowed.

CHAPTER 3

WHAT TO DO ON FINDING A FIRE

Raise the Alarm!

Fig 3-1. Raising the Alarm

0301. Raise the alarm with loud shouts of '**FIRE FIRE FIRE**', giving the location of the fire. If a ship's internal telephone is at hand dial HQ1/QM (999) and report the fire giving the location, size and type of fire, details of any casualties and your name, as the person who found the fire.

0302. If the Fire is Behind Closed Doors

If smoke is seen coming from behind a closed door it gives no indication to the size or type of fire behind it. **Do not tackle.** Obtain firefighting equipment in preparation for fighting the fire. Brief the Attack Party leader then the LMEM(M) i/c the Main Group, he will then provide you with an escort to take you to the OOD in HQ1/SCC.

0303. If the Door of the Compartment on Fire is Open

Raise the alarm as illustrated above and then carry on as follows:

a. If you are fully clothed attack the fire using first aid appliances appropriate to the type of fire. Remember, **keep low** and make sure you have a safe line of retreat. Brief the Attack Party leader, then the LMEM(M) i/c the Main Group, he will then provide you with an escort to take you to the OOD in HQ1/SCC.

Fig 3-2. Moving a Casualty

b. If, after your initial attack you are beaten back **close the door to the space** and start preparing hoses etc, as appropriate, ready for the Attack Party. Brief the Attack Party leader, then the LMEM(M) i/c the main group, he will then provide you with an escort to take you to the OOD in HQ1/SCC.

c. Move casualties to a safe place clear of the firefighting.

d. Consider whether the fire can be prevented from spreading. If possible carry out appropriate action to contain. This can be achieved by cooling the boundary of the fire; removing flammable articles from the path of the fire; stopping local ventilation and switching off local electrical circuits.

e. Stay in the vicinity of the fire ready to brief whoever is in charge, as follows:

(1) Where the fire is situated;

(2) What is burning;

(3) Any hazardous circumstances;

(4) Position(s) and number(s) of casualties;

(5) Achievements regarding the preparations for fighting the fire, eg boundary cooling, equipment to hand or rigged etc; and

(6) Details of electrical circuits and of any local ventilation fans which have been switched off.

f. On completion of e. above go to HQ1/SCC and brief the OOD or senior rating in charge, with the escort provided by the person in charge at the scene of the fire.

CHAPTER 4

FIREFIGHTING ORGANIZATION

0401. Introduction

A fire can occur at anytime during a ship's service - in harbour or at sea - in peacetime or in wartime. The organization developed to deal with such a situation is explained in this chapter.

0402. A ship's organization and firefighting procedures must be capable of dealing with the worst possible case - a major fire when only a minimum number of hands are available to deal with it.

0403. Standard Harbour Fire Organization

The duty part of the watch in harbour (whether alongside, outside a Fleet base or at anchor) is the smallest unit that might be required to fight a major outbreak of fire. Therefore, this body of men must be organized to provide the following:

a. **Specialist Personnel** to maintain overall control of the incident and safety of the ship.

b. **A Main Group** to locate, control and extinguish the fire as quickly as possible.

c. **A Containment Party** to establish the boundaries of the fire and to contain it.

0404. The number of hands in each group, detailed in the following paragraphs, are the recommended minimum (based on the complement of a fully manned frigate). If they cannot be met the risk of ineffective firefighting will be increased. Whenever it is necessary to make reductions in numbers it is most important to retain the overall concept of the organization.

0405. To promote the effectiveness of this fire and emergency party the following actions are recommended:

a. Daily orders should include the names of all members of the fire and emergency party and detail their specific duties within the organization.

b. At the forenoon muster of the fire and emergency party, or on taking over the duty, all the members are to be fully briefed on their individual tasks and responsibilities. It is too late to attend to these arrangements once an emergency has arisen.

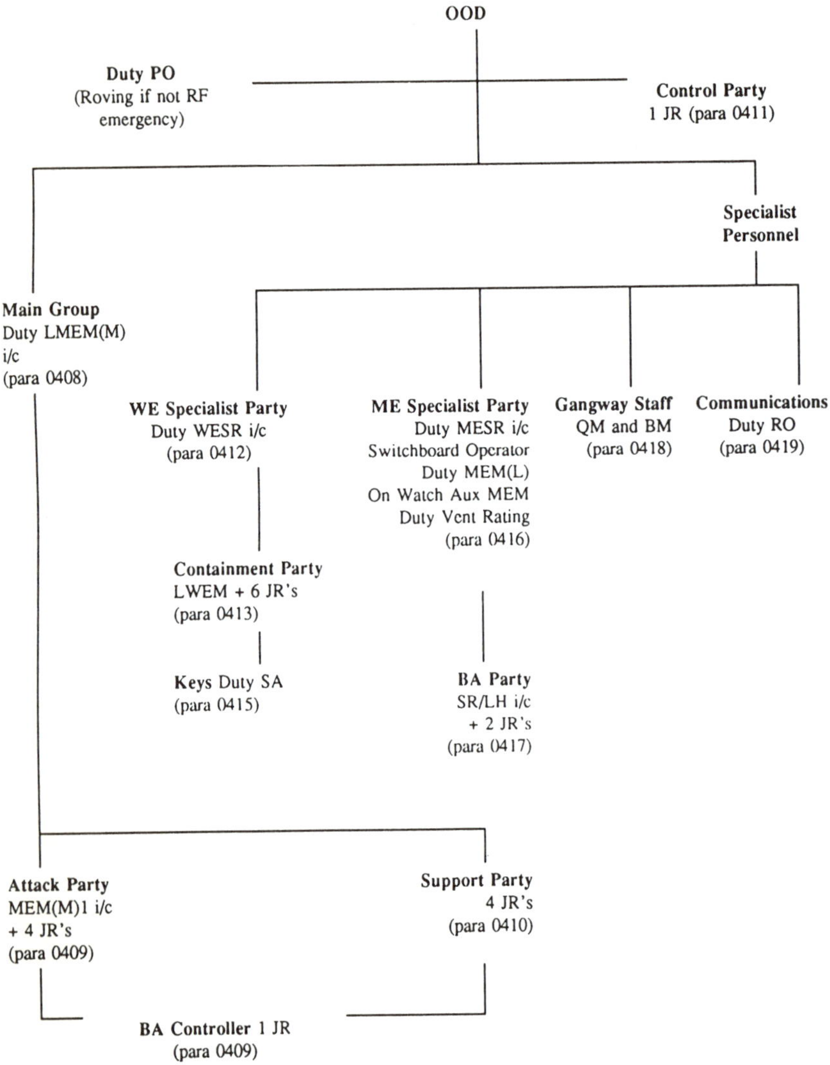

Fig 4-1. Standard Harbour Fire Party Organization

Notes:

*1. The **Main Group** should be under the overall charge of the Duty LMEM(M) who must have a thorough knowledge of the ship and firefighting techniques.*

*2. The **Attack Party** consists of an MEM(M)1 in charge plus at least one other MEM(M), one WEM, one Operations Department junior rating and one other JR. They are to be dressed in basic firefighting rig throughout the period of their duty. Two members of the attack party are to be BA wearers dressed in basic firefighting rig and BA.*

*3. The **Support Party** should consist of all those ratings not specifically detailed for either the Attack or Specialist Parties and should be under the charge of the 2nd duty MESR (No 3 of the firefighting team) or, if he is not available, a suitably experienced MEM.*

*4. The rating in charge of the **Breathing Apparatus Party** is the BA co-ordinator and must be specifically trained in this task. All members of this party must be familiar with BA control and recharging techniques and the use of the BA TOTE BOARD.*

*5. The **Containment Party** is to be led by the Duty LWEM. He is to liaise directly with the Duty WESR who is to be the containment co-ordinator within the control group at HQ1.*

6. The Duty LWEM will also provide immediate guidance on electrical and electronic safety and implement technical advice from the Duty WESR.

7. The containment party should contain all other duty personnel who have not been detailed for other specific fire and emergency tasks. Special duties ratings such as cooks, stewards and flight deck ratings should not be forgotten.

8. In the full harbour situation all ships should employ an ME senior rating and a separate qualified switchboard operator.

9. When wearing BASCCA personnel with beards or small faces experience a serious reduction in wear time due to the reduced efficiency of the face seal. When practicable, these personnel should not be allocated duties which require them to wear BA. (In wartime all beards will have been shaved off.)

10. Attack and Support Party share the BA Controller.

12. Duty SA musters at containment control.

13. BA party is under DMESR.

14. BM can leave the gangway to rig guideline as directed by OOD.

15. Personnel to include Duty Watch First Aid Qualified Rating (Non RF).

0406. First Aid Firefighters

These are the people who discover a fire and then carry out the drill in Chapter 3, WHAT TO DO ON FINDING A FIRE. **Remember** the person that found the fire must be sent to HQ1 with an escort, as soon as they can be spared or when ordered by the OOD.

MAIN GROUP

0407. The precise function of the Main Group may alter depending upon the situation and nature of the fire, but the basic philosophy remains the same - the Attack Party must always be prepared to attack a fire **immediately** it is reported and the remainder of the group then maintains this attack **continuously** and **aggressively**. The Main Group is led by the Duty LMEM(M) who is identified by wearing a red surcoat with a white stripe.

0408. Duty LMEM(M) Guidelines and Instructions

a. He should have read and understood the following publications:

(1) BR 2170 Volume 1, Chapters 30 and 31.

(2) Ship's Standing Orders.

(3) Ship's NBCD Standing Orders.

Individual ships may differ as to where the details of their firefighting orders and instructions are laid down

(4) Harbour Fire Exercise Assessment Form (paying particular attention to Umpire 1 section).

b. On taking up his duty he is to muster the duty watch to ensure that they are aware of their tasks and responsibilities within the fire and emergency party organization.

c. On hearing the alarm he should proceed immediately to a safe position as close as possible to the scene of the fire, in order to appraise the situation and to ensure that an effective initial attack is being made on the fire, and inform the OOD in HQ/SCC of his whereabouts. With this first hand knowledge, and after debriefing the person who found the fire, he will be able to decide where to position himself and where to establish, after consultation with the DWESR, the smoke boundary.

d. He must advise HQ1 when the fixed firefighting systems should be used.

e. He must also bear in mind that the attack BA men:

(1) Should be dressed in overalls/No 8 working dress with sleeves rolled down.

(2) Should carry out the 10 second face seal check before entering the fire zone.

(3) Will not necessarily be relieved by the four man firefighting team as they may have either put the fire out, run out of air or been beaten back by the fire.

(4) Will have been nominated at the duty watch muster so that the BA Controllers' boards (Form S3047) can be made up in advance. (In making up the boards it is to be assumed that the pressure in each BA is 180 bar, and the time started to breathe taken as time the incident was piped.)

(5) May be compelled, for any of a variety of reasons, to withdraw from their attack on the fire. Should this happen they must ensure that the compartment is sealed by shutting the hatch or door. Consideration may also be given to holding the hatch or door on a waterwall, should a waterwall nozzle be available and the effect on ship stability acceptable.

f. The Duty LMEM(M) must ascertain if there are any men trapped in the fire zone and if there are any injured men who require assistance.

g. On the withdrawal of the attack BA men, the Duty LMEM(M) must ensure that the used firefighting appliances are moved out of the area from which the next attack is to be mounted.

h. He must request, from HQ1, information on any specific hazards existing in the compartment on fire or in adjacent compartments. This information is included on form S3021, Compartment Killcards, which are held at HQ1.

i. He must supervise hose running, detailing the lengths of hoses required and the type of nozzles to be used. He should be guided by the following principles:

(1) Firefighting hoses must be run from separate hydrants though it is acceptable to use a breeching **Y** piece to supply a boundary cooling hose and a firefighting hose from the same hydrant.

(2) Sufficient hose must be provided to allow the firefighters to reach all parts of a compartment.

(3) The choice of portable nozzles is limited to the Firefighter nozzle (for use by the Fire Fighter) and the Waterwall nozzle (for use by the Waterwall operator.) The Firefighter nozzle has a considerably lower discharge rate than the Waterwall nozzle and is therefore the one used for firefighting between decks.

(4) Waterwall nozzles must always be used to provide the waterwall, and the Firefighting nozzle used for fighting carbonaceous fires.

j. He must establish the nature of the fire and if there is a possibility that it is a fuel/oil fire he must ensure:

(1) The firefighting medium is AFFF.

(2) An FB5(X) branchpipe is used, in conjunction with an inline inductor where appropriate.

(3) Sufficient quantities of replenishment AFFF are provided for the support party. A 20 litre drum lasts approximately sixty seconds when supplying an FB5(X) branchpipe through a No 2 hose.

(4) The number of drums to adequately blanket a compartment is shown adjacent to the foam inlet tube and this number should not be exceeded. It must be remembered that the foam will not spread evenly throughout a compartment owing to obstructions such as machinery, lockers, consoles etc, or if the ship has a heel.

k. As soon as the Support Party arrives (and while the Attack Party is still fighting the fire) the Duty LMEM(M) must brief them on the nature, location and size of the fire together with other information available. This can be taken from Killcards and should include positions of hazardous items such as gas bottles, batteries etc. This briefing should take place in a suitable clean air environment where the team need not use air from their sets. He must ensure that each member of the team is conversant with his task within the team.

l. The Duty LMEM(M) must ensure that all personnel not involved with a re-entry procedure are kept well clear of the door or hatch to be used. Having established this personnel boundary he must ensure that all those remaining in this area keep low to avoid any fireball.

m. As soon as the Support Party has been deployed to fight the fire the Duty LMEM(M) must ensure that the relieved Attack Party is being dressed and prepared to form a second Support Party to be ready for briefing well before the team tackling the fire is due out.

n. He will debrief all firefighting teams on their exit from the fire zone.

o. He must keep the control team informed of the current situation especially when it is changing, for better or for worse. In the latter situation requests for additional manpower should be made early.

p. The Duty LMEM(M) must continue to direct the main group's attack on the fire until he is sure that the Local Authority Fire Brigade (LAFB) (which should arrive at the scene of the fire (SOF) quickly in the UK, and probably less quickly in some other parts of the world) is ready in all respects to commence firefighting and has been fully briefed on taking over the firefighting task. He remains at the SOF and:

(1) Liaises with the LAFB.

(2) Maintains control over RN personnel who are assisting the LAFB in carrying out other firefighting tasks.

(3) Keeps the OOD informed of the current situation and so assist him in his overall ship safety function.

0409. The Attack Party

The Attack Party consists of an MEM(M)1 in charge plus four other junior ratings all of whom are to be dressed in basic firefighting rig - No 8 working dress or cotton overalls, woollen socks, DMS boots and red surcoats. Their aim is to attack the fire as quickly as possible.

a. On the alarm being raised the MEM(M) plus two previously detailed junior rates are to attack the fire collecting SPE(AFFF) and CO_2 fire extinguishers en route. They should establish the seat and nature of the fire using the Thermal Imaging Camera (TIC) if it has been delivered by the nominated rating of the Containment Party.

b. The two men detailed as BA wearers, on hearing the alarm, are to don BA and proceed to attack the fire whether or not the BA Controller is present. The men being relieved (MEM(M) + 2) who are not wearing BA **must then withdraw** (see sub-para d). The effective use of BA will markedly improve the chances of maintaining a **continuous attack** on the fire.

c. The BA controller will obtain his own BA control board (Form S3047) and enter on it the time of the fire pipe as the time of entry of the attack BA team. He may not see the Attack BA men prior to them making their entry and must assume their initial BA pressure as entered on the control board to be 180 bar.

d. When relieved, or beaten back and their attack on the fire has to cease, the Attack Party is to join up as the second Support Party and prepare for further firefighting.

Notes:

*1. **Thermal Imaging Camera** (TIC). The prime use of the TIC is to locate the seat of the fire through smoke. It is also useful in search and rescue, identifying fuel leaks feeding a fire, locating firefighting equipment and spotting unexpected hazards in smoke logged areas. Until the arrival and involvement of the fearnought suitmen, the TIC is a valuable tool to the MEM(M) of the Attack Party. Initially the TIC is to be given to the MEM(M) i/c of the Attack Party by a member of the Containment Party (see para 0413b). Subsequently its use will be controlled by the LMEM(M) i/c of the Main Group. In most circumstances where the TIC is needed the user will be wearing BA.*

*2. **Large Compartment with More Than One Access Point.** There is a risk that when the alarm is raised members of the Attack Party will proceed to differing access points and frustrate the aim of a co-ordinated* *attack. Provision is to be made in Ship's Orders for a standard muster point for each major space, so that these become well known and can also be piped when the initial alarm is raised. Simplicity is vital, therefore choice of Port or Starboard passage access and fwd or aft door to all compartments may be appropriate as standard operational procedure.*

3. ***BA Controllers.*** *All BA controllers are to be in possession of a watch for the period of their duty. Their duties are:*

a. To ensure that BA users are dressed correctly before entry to a fire.

b. To control BA users before entry to, and after exit from, a fire.

c. To be fully competent in the use of a BA control board, (see also para 0718), and the drill for BA face seal checks.

0410. Support Party

On the alarm being raised the Support Party is to muster, with all speed, at the designated FRPP. This party has two main tasks:

a. To provide men dressed in full firefighting rig, (see para 0703), with a BA controller, to the scene of the fire as soon as possible.

b. To co-ordinate, collect and provide to the SOF all necessary equipment to sustain a continuous attack on the fire (hoses, nozzles, inline inductors, AFFF, etc). Manpower for the latter task is to be provided from those remaining in the Support Party when the first task is achieved, members of the retiring Attack Party and from supplementary personnel onboard when the alarm is raised.

Support Party fearnought suitmen are to be dressed in basic firefighting rig throughout the period of their duty. Surcoats are to be worn (see para 0422). On the alarm being sounded these men are to don anti-flash gloves and anti-flash hood, seamen's stockings, firefighting boots, fearnought suit and firefighting gloves, BA, bulb horn, protective helmet with portable radio communications set, (transceiver) and cap lamp as advised in para 0424. When fully dressed they are to be taken to the SOF by the Team Leader and reported to the LMEM(M) i/c of the Main Group and the BA controller. The support party should be deployed, initially, as follows:

- **Waterwall** operator.
- **Firefighter** with the appropriate appliance.
- **Team Leader** with Thermal Imaging Camera.
- **Hose Tender**
- **Ship Protector** (if required).
- **BA Controller** Controls both the Attack Party and the Support Party.

CONTROL AND SPECIALIST PARTIES

0411. Control Party

On the alarm being raised, the control rating is to close up in HQ1 (or another suitable position with adequate communications should it not be possible to occupy HQ1) and man the incident board and internal communication lines. He is to make a chronological record of events during the incident. The OOD must not become too involved in the details which are the responsibility of the ratings in charge of the various groups and parties, but he must consider the following:

a. The possible hazards from explosives and flammables.

b. The availability of equipment, power supplies, fire and salvage pumps.

c. The hazards of smoke and smoke clearance.

d. Safety of personnel, including any civilians who may be onboard.

e. Ship stability.

f. Control and Allocation of additional manpower.

g. Liaison with adjacent ships and outside authorities.

Notes:

1. Smoke clearance action must neither fan the fire nor smoke-log otherwise clear areas crucial to the firefighting task. There are two main methods of clearing smoke.

a. Using large exhaust fans such as the galley and main machinery space fans to suck the smoke from the affected areas.

b. Using large supply fans to build up a pressure to blow the smoke from the ship.

2. In both cases (notes 1a and 1b above) there will normally be a need to open up an air route along which smoke can pass to atmosphere, or to facilitate the required air flow to achieve effective smoke clearance.

3. Smoke control and clearance exercises should be practised regularly so that the necessary personnel are made familiar with the arrangements and equipment fitted.

0412. WE Specialist Party

On the alarm being raised, the duty WE Senior Rate is to close up in HQ1. He is responsible for:

a. Ensuring firefighting teams have access to magazines in the vicinity of the fire.

b. Advising the OOD on the following:

(1) Magazine and explosive safety.

(2) Movement of ammunition.

(3) Magazine fire fighting arrangements, spray systems and fitted pumping arrangements.

(4) Electronic and electrical equipment safety.

(5) Internal and external communications.

c. Acting as Containment Co-ordinator, liaising between the rating in charge of the containment party, specialist parties and the OOD, in order to establish priorities for the containment task and to ensure that the duties listed in para 0413 are effectively executed.

0413. Containment Party

On the alarm being raised, members of the Containment Party are to muster with all speed at a convenient point near to, but separate from, the Support Party. The location of the muster point must be included in the initial fire alarm broadcast. The duties of the Containment Party include:

a. Establishing the immediate boundary of the fire by searching all adjacent compartments, lockers and stowages - not forgetting ventilation trunking to and from the affected area.

b. Provision of a TIC to the SOF. A rating is to be predesignated to ensure this happens.

c. Removing flammable materials from the vicinity of the fire boundary.

d. Action to limit the spread of smoke and preparations for smoke removal. Specific instructions are to be obtained from HQ1 before affecting smoke removal measures.

e. Rigging of boundary cooling, monitoring of bulkhead temperatures and provision of manpower to implement boundary cooling where necessary.

f. Action to minimise the dangers to personnel from electrical equipments and to prevent, as far as possible, damage to electronic and electrical equipments from the effects of firefighting/boundary cooling water.

g. Action to remove or drain down free surface water.

0414. The Duty LWEM, as leader of the Containment Party, is to be briefed by the Duty WESR on the location of all the adjacent compartments and to be made aware of those compartments presenting the greatest hazard, eg, magazines, gas bottle stowages, fuels, flammables, etc. He should constantly check the boundaries to ensure that his team has contained the incident and is to keep the OOD informed through direct liaison with the DWESR. He is to provide first hand guidance on the safety of electronic and electrical equipment and personnel. Smoke control measures initiated by the Attack Party become the responsibility of the Containment Party once it has closed up. He is to initiate local smoke control measures, using ventilation, as directed by HQ1, and to report back on their effectiveness and the location of smoke boundaries.

0415. Keys Rating

On the alarm being raised, the rating responsible for keys is to obtain the keys to the duplicate keyboard from the OOD, and then provide to the ratings in charge of both Main Group and Containment Party the necessary keys to enable firefighters to gain access to all compartments in the vicinity of the fire. He then returns to the containment control point.

0416. ME Specialist Party

This party is responsible for:

a. Maintaining the integrity of the ship's electrical power supplies and safeguarding those to electrical services, if necessary by shedding load and/or putting additional generators on load. Where possible, normal and alternative power supplies to the fire pumps should be established and maintained during firefighting operations.

b. Crash stopping ventilation, shutting upper deck vent flaps on systems serving the compartment on fire, and subsequent operation of other vent systems for smoke clearance.

c. Checking all fire pumps and starting additional pumps to maintain the high pressure sea water (HPSW) system pressure. If necessary, this is to be achieved by shutting off sea water to non essential services.

d. Isolating electrical power supplies in and around the area of the fire to minimize the danger to firefighters and equipments.

e. Checking and safeguarding those systems essential to the firefighting effort (saturated steam, HP air, etc) and isolating those systems containing hazardous fluids (fuel, lub oil, hydraulics, etc) to prevent them from feeding the fire.

f. Making contact and liaising with the shore electrical supply authority. (When in harbour.)

g. Advising the OOD on fixed firefighting installations for machinery spaces, paint shop, flammable store rooms, gas bottle stowages, hangars, etc, and, if appropriate, starting auxiliary boilers to supply saturated steam.

h. Advising the OOD on the following:

(1) The technical aspects of firefighting.

(2) Pumping and salvage arrangements.

(3) Ventilation and smoke clearance.

(4) Ship stability.

(5) The presence of compressed gases or pressure vessels in the vicinity of the fire, including HP air, refrigerant gases, oxygen and acetylene gases, beer kegs and associated CO_2 bottles.

(6) The nature of the ship's structure, ie, aluminium or steel on the area of the fire.

0417. Breathing Apparatus Party

On the alarm being raised the BA co-ordinator organizes his party to collect BA and spare cylinders to form a dump in a position allowing room to work and free from smoke. This location is to be broadcast. Once the necessary BA has been collected (excluding the 4 designated for the upper deck re-entry locker) the party becomes responsible for:

a. Provision of BA to required user locations.

b. Recharging used BA cylinders.

c. Informing the OOD of the BA availability state (Using the BA Co-ordination (Tote) Board.)

d. Co-ordinate optimum use of BA procured from outside sources (other ships etc).

e. Provision and briefing of other BA controllers other than BA controllers predesignated as members of the main group. (See Note 2 below.)

Notes:

1. Both the BA co-ordinators and the ratings detailed to recharge cylinders are to be trained in recharging techniques.

2. The BA Controller in the Main Group (Attack Party) is not a member of the BA party as such.

3. The BA co-ordinator is responsible for ensuring that all men detailed for BA control duties are fully briefed at the start of their period of duty.

4. Careful consideration must be given to the location of the initial, and other BA dumps. Ideally, when smoke logging dictates the need to move the dump away from the vicinity of the activated FRP an alterative position should be selected on the upper deck to SOF guideline route. This ensures that personnel wearing BA will be able to locate the BA dump easily on withdrawal and that the ship's and the LAFB dumps can be collocated. As standard operating procedure (SOP) the LAFB set up their BA dump on the upper deck to SOF guideline route.

5. Fire produces formidable smoke and toxic fume hazards. BA co-ordination and BA wearer control are essential and vital parts of the firefighting organization and must be practised regularly.

0418. Gangway Staff

On receipt of the fire alarm, from whatever source, a pipe is to be made by the gangway staff or HQ1/SCC watchkeeper, as appropriate, informing the ship's company of the existence of a fire, its location and where various members of the fire and emergency party are to muster - this pipe is to be repeated. When available the main broadcast alarm is to be used prior to making this initial emergency broadcast. Thereafter the gangway staff or the HQ1 watchkeeper (depending on access to shore telephone) are to carry out the following:

a. Inform the Local Authority Fire Brigade (LAFB).

b. Inform shore authorities.

c. Inform adjacent ships.

d. If the LAFB has arrived then once the appropriate route to the SOF from the **ON** brow has been agreed between the officer responsible for ship safety and the LAFB officer the ships staff are to identify the route by running a combined guide and sound powered telephone line.

e. Control entry to and exit from the ship.

f. Prepare to meet the LAFB Officers and escort them to HQ1/OOD.

g. Keep a red flag in daytime, or a red light at night at the foot of the **ON** brow.

0419. Communications Rating

On the alarm being raised, a Duty RO is to carry out the following tasks:

a. Rig a sound powered telephone between HQ1 (or other control position) and the SOF.

b. Man emergency communications nets to outside authorities.

Notes:

1. Task a above should be undertaken by the HQ1 incident board operator/ communications number when in the reduced harbour fire party organization.

2. Task b is only to be undertaken when the full fire party organization is in force and there are two duty RO's in the duty watch.

EMERGENCIES

0420. As soon as it becomes apparent that a fire is more than a minor incident, and that the Attack Party are not extinguishing the fire, the OOD should not hesitate to declare a full scale emergency. It will ensure that all uninvolved personnel are brought out of the ship and clear of danger. It will also provide a ready pool of spare hands for extra firefighting and containment tasks. The correct pipe for initiating an emergency is **'Emergency, Emergency, Emergency, Clear lower deck of all personnel not involved in firefighting, muster on the ..... keeping clear of the fire area'.** (..... pipe should give the appropriate muster point bearing in mind the weather, do not always use the flightdeck as a matter of course.) This broadcast should be prefixed by the use of the main broadcast alarm.

PERSONNEL SAFETY AND CONTROL

0421. In the event of a full scale emergency being declared, it is important that every person onboard (including civilian personnel where appropriate) can be accounted for as quickly as possible. Ratings in charge of the various duty groups and parties should account for their personnel and report to the OOD. Departmental nominal lists should be compared with leave boards/cards to ascertain the names of those who should be onboard and an officer or rating should be nominated from available personnel to muster non duty personnel and report back to the OOD.

IDENTIFICATION OF PERSONNEL (FIRE AND EMERGENCY PARTY)

0422. In order to facilitate the easy recognition of fire and emergency party personnel coloured surcoats should be worn as follows:

- Red with a white stripe - rating i/c Main Group.
- Red - Main Group with the exception of the BA controller.
- Yellow - BA party, plus the BA controllers of the Attack Party and the Support Party.

USE OF IMPROVED FIREFIGHTING COMMUNICATION SYSTEM

0423. Five portable radio transmitter/receivers (transceivers) will be available in most ships, at each FRPP. Each set consists of a lightweight protective helmet (Slim Tank Helmet (STH)) including microphone and earphones, the portable transceiver and press to transmit switch.

0424. These sets are deployed as follows:

- Support Party fearnought suitman - Waterwall.
- Support Party fearnought suitman - Firefighter.
- Support Party fearnought suitman - Team Leader.

Firefighting organization

- Support Party fearnought suitman - Hose Tender.
- Rating i/c main group (provided by support party BA controller).
- The sets from other FRPP's may be used by the reliefs of the Support Party (until the full proposed allowance of 10 sets per FRPP is achieved).

0425 - 0430. Spare

REDUCED HARBOUR FIRE ORGANIZATION

0431. When a ship is in a Fleet base, and expert firefighting assistance is readily available from its LAFB, the size of the fire and emergency party may be reduced, but not in the specialist parties, in whose duties the LAFB are not trained. In some ships, the availability of Deterrence and Response Forces (D&RF) trained manpower, particularly junior ratings, may not permit the Attack Party to be divorced from security duties. If unavoidable, this can be accepted but only in the reduced harbour fire organisation.

0432. This reduced organization is independent of the Notice for Sea and represents the minimum number of people who can be retained onboard whenever a ship is operational. It should not be further reduced. However, when ships are working a dormant leave routine, the fire and emergency party can cover more than one ship and its numbers can be found from the total duty personnel of **paired** ships.

0433. When a ship is non operational the problems are somewhat different and the composition of the Fire and Emergency Party when the ship is in refit is not covered here. (***Note.*** *Ships in DED are usually still operational.*) However, when the composition of a fire and emergency party is less than twenty the following guidelines apply:

a. The OOD's position and fire dump is to be established in a specified area, eg, the hangar.

b. An Attack Party with ratings detailed to wear BA is nominated, and they must be prepared to don full Fearnought if the need arises.

c. Any additional personnel on board must form a Containment Party.

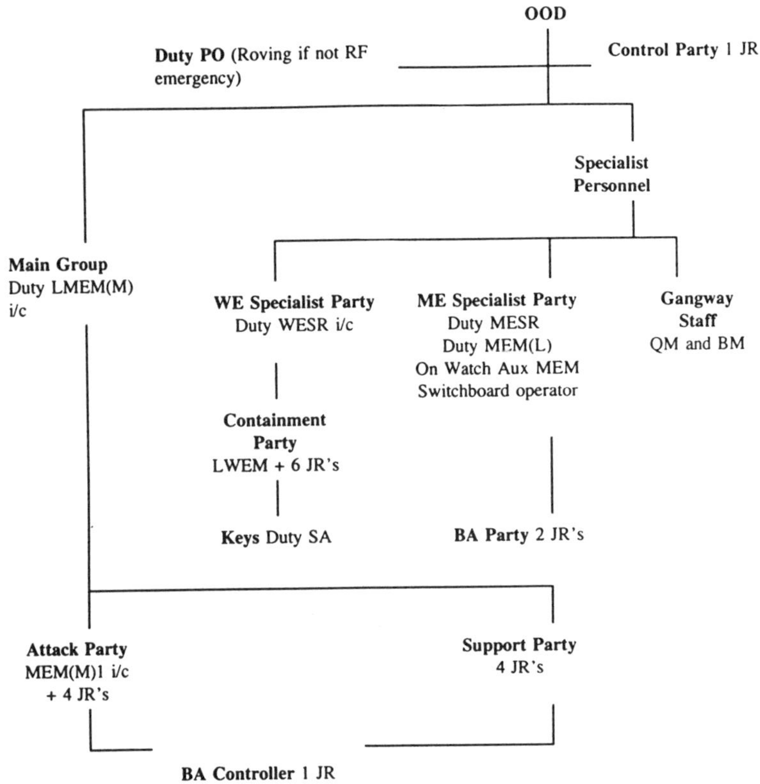

Fig 4-2. Reduced Fire Party Organization

Notes:

1. The Attack Party under the leadership of an MEM(M)1 should contain at least one other MEM(M), one WEM, one Operations Department junior rating and one other JR. They are to be dressed in basic firefighting rig throughout the period of their duty and must be able to hear all pipes ie within the range of main broadcast speakers (and not wearing "Personel Stereos" in order that they may respond as quickly as possible.

2. The four members of the Support Party are to dress in full firefighting rig and BA before reporting to the rating in charge of the Main Group. The Attack Party BA controller is to act as controller for all members of the Main Group.

3. In addition to taking charge of the Containment Party the LWEM should also provide guidance on electronic and electrical safety.

4. The Containment Party should contain all other duty personnel who have not been detailed for other specific fire and emergency party tasks. Special duties ratings such as cooks, stewards and flight ratings should not be forgotten.

5. The on watch Bosun's Mate may run out the guideline from the gangway to the SOF. (see 0418d).

6. The duty SA is to act as a member of the Containment Party once he has obtained the keys for all compartments in the fire area.

7. Provided he is suitably qualified, the duty ME senior rate may, in ships where the main switchboard and HQ1 are co-located, also undertake the duties of switchboard operator.

FIREFIGHTING AT SEA - STANDING SEA EMERGENCY PARTY (SSEP)

0434. In all ships at sea there must be a party of specifically detailed firefighters under the charge of a suitable senior rating, which is available at all times and ready for all immediate response to a fire. The party should be made up from daywork personnel.

0435. Organization

The Standing Sea Emergency Party (SSEP) should comprise all the firefighting groups and parties detailed in the previous paragraphs (Main, Containment and Specialist), when daywork manpower is readily available.

0436. In smaller ships the number of daywork hands available to form the SSEP is likely to be more limited and therefore, in frigates and destroyers, the suggested minimum requirement is for an Attack Party comprising one senior rating in charge of six junior ratings and, in addition, the necessary control personnel. All the ratings should be daywork personnel and the aim should be to achieve continuity and a well trained team. Although the actual composition of this party is at the Commanding Officers discretion, the most suitable Attack Party senior rating is probably the CMEM(M). The junior ratings should ideally be drawn from all departments and one should be from the Ventilation Party. They should not be detailed as Special Sea Dutymen.

0437. In ships smaller than a frigate, or those where daywork manpower is very restricted, the size of the SSEP may need to be further reduced. If this is the case, the concept of the Attack Party should still be retained. For example, where there are only four junior ratings available to form the attack element, two are to don BA.

0438. Emergency Stations

If the SSEP is unable to cope with a fire, either due to lack of personnel or shortage of material resources, it is important that the ship be piped to emergency station in good time. This will get the right people to the right places quickly and bring the ship's Fire and Repair Parties to State 1. It will also clear the area of unwanted personnel, allow off watch watchkeepers to be accounted for and provide back-up where necessary. The actual stage at which the ship should be piped to Emergency Stations will obviously vary depending upon the size of the SSEP and the rate of escalation of the incident.

0439. Identification of Personnel (Standing Sea Emergency Party)

In order to facilitate the identification of SSEP personnel coloured surcoats should be worn as follows:

- Red with white stripe - Senior rating when i/c of the SSEP.
- Red - Attack Party and main groups of ships larger than DD/FF with the exception of BA controllers.
- Yellow - BA Party, plus the Attack Party and Support Party BA controllers.

0440. Procedure

In ships larger than destroyers or frigates where the SSEP comprises all the firefighting groups and parties of the full harbour organization, the procedure to be followed is much the same as when the ship is in harbour.

0441. When the alarm is raised in a ship with only an Attack Party of six junior ratings, the rating in charge and three others should proceed directly to the SOF. They are to collect portable first-aid firefighting appliances and the TIC en route, and join up with such firefighters as there may be. Having established the seat and nature of the fire, they are to fight it with the appropriate appliances. The two ratings detailed as BA men are to don their sets and proceed to attack the fire whether or not the BA Controller is present. The men being relieved, who are not wearing BA, **must then withdraw**. The BA Controller will obtain his own control board and enter on it the time of the fire pipe as the time of entry of the attack BA men and a pressure of BA as 180 bar. All members of the attack party are to be fully experienced in dressing and wearing BA as well as being capable of using a control board.

0422. In those ships with numbers other than six junior ratings in the Attack Party the firefighting procedure should be adjusted to suit the numbers involved.

0443 As soon as the initial alarm is raised, HQ1 should be manned by the Action NBCD Officer, an incident board operator, a communications number and the Explosives Safety Officer (ESO); Switchboards should be manned by qualified switchboard operators to crash stop ventilation and to safeguard electrical supplies; SCC/MCR/ER watchkeepers should put all available fire pumps on to the HPSW systems. The first task of the communications number is to rig a sound powered telephone between the control position and the SOF.

0444. The Executive Officer should briefly visit the SOF and HQ1 in the early stages, in order to gain an up to the minute appreciation of the situation so that he will be able to give advice to the Command on whether there is a need for the ship to go to emergency stations.

0445. First Aid Firefighting During Silent Hours

It is recommended that the following on-watch ratings be detailed to undertake first aid firefighting until relieved by the SSEP; an MEM, an RO, a Bosun's Mate and an Operations Department rating either from the Operations Room or the watch on deck.

0446 - 0450. Spare

FIREFIGHTING IN CONJUNCTION WITH LOCAL AUTHORITY FIRE BRIGADES

0451. Liaison

A good liaison with Local Authority Fire Brigades (LAFBs) is essential if efficient use is to be made of their service in the event of fire. This liaison is vital when a ship is in harbour for any length of time and the ship's company is depleted (eg by leave or training).

0452. Contact the LAFB in a home Naval Base should be made, in the first instance, through the Ministry of Defence Area Fire Prevention Officer (AFPO). The AFPO should be kept up to date with any significant changes that might affect the ship's firefighting capability, eg, docking, leave period, supply of additional pumps, etc. The AFPO or one of his officers will attend ships fires during normal working hours, but this attendance may be delayed during silent hours. It is therefore most important to establish a direct liaison with the LAFB.

0453. Successful liaison with the LAFB depends upon an understanding of any differences in aims between the RN and LAFB.

0454. The firefighting priorities of the LAFB are as follows:

a. To save life.

b. To save property.

c. To extinguish the fire.

0455. a. Ships Standing Orders are to contain instructions for calling the LAFB as an automatic response to the discovery of a suspected or known fire when a ship or submarine is berthed in a Naval Base, commercial port or shipyard. At anchor or at a buoy this assistance should be considered at an early stage, paying due regard for the ships geographical location and the probability of useful and timely assistance from local authorities.

b. Upon his arrival, the LAFB Senior Fire Officer attending the incident is to be met at the brow and escorted directly to the ship's officer responsible for the safety of the ship (Officer of the Day (OOD) for RN ships or submarines, Duty Deck Officer for the RFA). At this stage the LAFB may establish a Control Point (CP), Forward Control Point (FCP) and a breathing apparatus and firefighting equipment dump in the most appropriate locations. Should the LAFB subsequently become involved in the incident, communications are to be established between the officer responsible for ship safety and the LAFB Control Point and, whenever possible, an RN/RFA liaison officer or senior rating should be stationed at the LAFB Control Point.

c. The ship's officer responsible for the safety of the ship is to fully brief the Senior Fire Officer on the situation, and then, after consultation with him, decide on the appropriate strategy to deal with the incident. Dependent on the nature of the incident and the ship's resources available to him the ship's officer responsible for ship safety may then request the LAFB to undertake some or all aspects of the operation. Any request for assistance should be clearly recorded. If LAFB assistance is accepted control of the firefighting aspects of the incident is to be given to the Senior Fire Brigade Officer.

d. The level of involvement of the LAFB will be determined by the operational state of the vessel and the ships complement. As a general rule frigates and above that are operational may require less assistance than minor war vessels or submarines. Due to limited manpower resources, RFAs and ships under repair will generally require major assistance from the LAFB. The initial level of attendance is predetermined by individual local authorities and will normally be in the order of 4 appliances.

0456. Ship Safety

Notwithstanding the involvement of the LAFB in some or all aspects of the operation the Commanding Officer, or his designated representative, retains full overall responsibility for all firefighting measures and the safety of the ship, but not the control of LAFB personnel if they are committed to the firefighting task. It is imperative that no matter what firefighting strategy is adopted, until such time that the fire is extinguished a continued attack on the fire is maintained.

0457. Jetty Space

Wherever physical conditions permit, a clear space (7.5m radius) should always be kept clear near the foot of the brow; LAFB appliances require a space 10.5m by 5.0m for parking and operation, although certain special appliances may require somewhat larger area. The area is to be conspicuously marked **Fire Brigade - Keep Clear**. The Naval Base authorities will provide assistance in the supply of suitable notices.

0458. Brows

When a ship has more than one brow a clear indication is to be given as to which one should be used by the LAFB officer to come abroad. A red flag should be used in the daytime and a red light at night, to mark the **ON** brow. The LAFB officer is to be met at the brow by an officer or senior rating with authority to answer his questions.

0459. Probable Questions

The following are among the questions likely to be asked by the first senior LAFB officer to arrive at the scene:

a. Is anyone missing and/or injured?

b. Where is the fire?

c. What is burning?

d. Have electrical supplies to the ship/fire zone been isolated?

e. How long have the fires been burning?

f. How did the fire start?

g. What action is being taken by ship's staff?

h. What is the state of the ship's HPSW system?

i. Are there any dangerous compartments adjacent?

j. What is the state of the ship's fire and emergency party?

k. How critical is the ship's stability?

l. How much water has been pumped in?

m. Are timber docking shores affected?

n. Has a smoke boundary been set up? If so, where?

0460. Missing Personnel

The LAFB may well arrive before an accurate check has been made as to whether any personnel are unaccounted for. During the initial consultations between the LAFB and the ship's officer in charge of firefighting a decision as to whether search and rescue operations by the LAFB are required in preference to firefighting assistance may well have to be made.

0461. Electrical Supplies

The LAFB normally expect all electrical supplies to an installation on fire to be isolated. This is seldom practicable in a warship on fire. However, when there is a risk of voltage in excess of 440 volts the equipment should be isolated. It must be noted that attempts at maintaining a **keep alive** policy may be counter-productive when compared with the savings in damage through quick extinction of the fire.

0462. Route to The Fire

Once the route to the fire from the **ON** brow has been agreed between the officer responsible for ship safety and the LAFB officer, the ship's staff are to identify the route by running a combined guide and sound powered telephone line.

0463. Control of Firefighting Personnel

Strict control over the combined numbers of personnel involved in the firefighting operation will need to be exercised and therefore, all RN/RFA personnel not directly involved must keep clear of the ship. This will be achieved by sounding the main broadcast alarm and making the pipe "EMERGENCY, EMERGENCY, EMERGENCY, CLEAR LOWER DECK OF ALL NON DUTY PERSONNEL, MUSTER ON THE JETTY (or other nominated position) KEEPING CLEAR OF THE FIRE AREA". Entry and egress must be controlled by the gangway staff at the brows.

0464. Withdrawal of Personnel

If the LAFB is delegated the task of finding and/or fighting the fire, RN/RFA personnel will be gradually withdrawn from within the smoke boundary as they are replaced by LAFB personnel. Ship's firefighters working in pairs and wearing BA, will normally be required to act as guides. Close collaboration between the officer or rating in charge of the ship's Main Group and the LAFB officer at the FCP is essential.

0465. The officer or rating in charge of the ships Main Group, once relieved by the LAFB at the scene of the fire, is to move to the FCP. His duties will be:

a. To liaise with the LAFB officer in charge at that position;

b. To keep HQ1 informed of all actions being taken;

c. To maintain control of all the RN/RFA personnel actively engaged in firefighting (but see para 0469).

0466. Evacuation of HQ1

Should HQ1 (or alternative position) need to be evacuated for any reason, it is recommended that the ship's control team moves to the brow nearest the LAFB control. If this is not practicable, any other location in clear air with good ship's internal communications should be selected (eg bridge wing).

0467. Communications

The ship is to provide a communication link from HQ1 (or alternative position) to the SOF, and from the brow to the SOF, manned by ships staff at both ends. LAFB officers will normally use their own communication systems, but these may prove inadequate in a warship environment and, in some cases, the associated RADHAZ prohibits their use.

0468. Use of Breathing Apparatus and Control

When the ship's firefighters are using BA, consideration should be given to establishing the BA co-ordinator at the FCP as soon as possible. The BA controllers (for members of the ship's fire party wearing BA) will then retreat up the guideline as the smoke boundary spreads. This will ensure that men wearing BA will find their controller when they leave the smoke boundary.

Firefighting organization

0469. Ship's staff BA controllers should continue to control ship's personnel using BA, at the same time maintaining the closest possible liaison with the LAFB controllers. Should firefighting measures be assigned to the LAFB, overall co-ordination of all BA wearers is to be exercised by the LAFB officer in charge of the firefighting operations.

0470 - 0474. Spare.

FIREFIGHTING IN WARTIME (STATES 1 OR 2)

0475. Firefighting in State 1

The local control of firefighting is undertaken by the FRP leader. The four man firefighting team (dressed in full firefighting rig) are used as the Support Party for men in the fire zone already attacking the fire. The remaining hands are used to form the relief firefighting team and in the BA support and containment roles. Additional support can always be sought from other FRPs and other positions as detailed and co-ordinated by HQ1, who should respond in direct relation to the size of the fire, damage and associated problems. See Figs 4-3 and 4-4 for the organizational diagrams.

0476. Firefighting in State 2

All the positions shown in Figs 4-3 and 4-4 are manned, but not necessarily by the personnel detailed in State 1. When these positions are manned in State 2 it is referred to as being in Defence Watches, and the numbers of men so stationed will vary, according to the complement and other considerations, from ship to ship. In the event of fire the men on watch attack as detailed in Chapter 3, the ship assuming State 1 immediately in order to get all the DC and FF personnel in position so that full firefighting teams can be organized and co-ordinated.

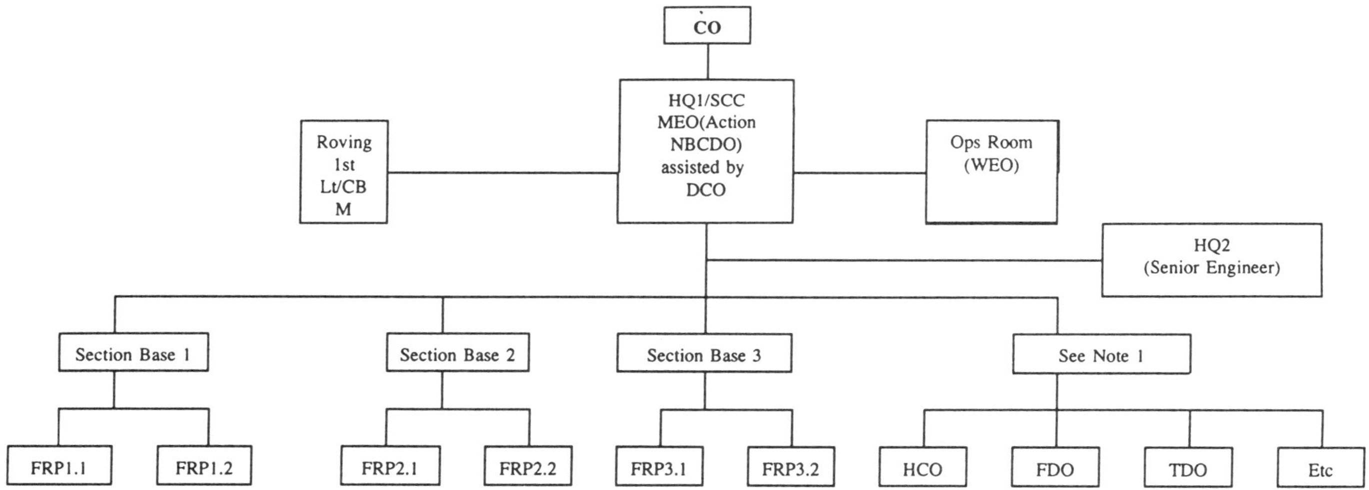

Notes

1. Areas of the ship concerned with specialized tasks, e.g., hangars, flight decks, tank decks, etc., are organized at section base level for control and reporting.

2. Rig for action stations; No 8s, DMS boots, anti-flash (hood and gloves) and action coveralls.

3. Each FRP details and dresses 4 men in full firefighting clothing, BA worn but unmasked, helmets to hand. Four other men are detailed as relief firefighters. Each team of four has a BA controller detailed.

Fig 4-3. Firefighting Organization (State 1) - Large Ships (Emergency Stations are Similar)

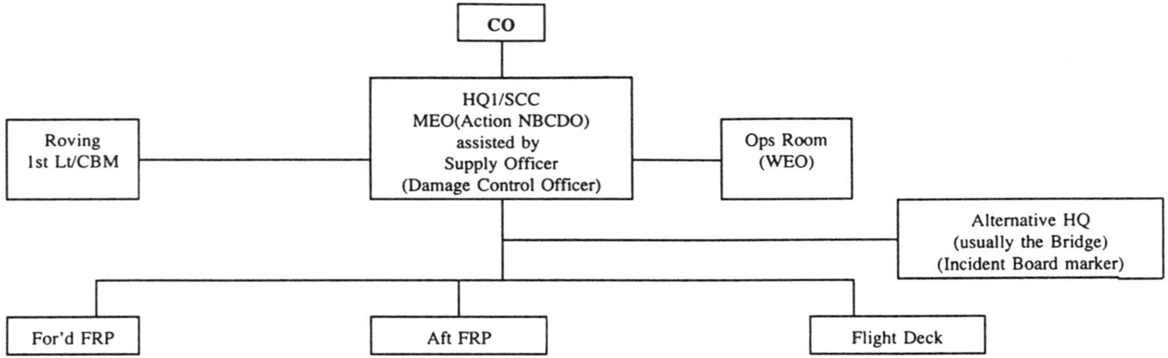

Notes

1. Rig for action stations; No 8's, DMS boots, anti-flash(hood and gloves) and action coveralls.

2. Each FRP details and dresses 4 men in full firefighting clothing, BA worn but unmasked, helmets to hand. Four other men are detailed as relief firefighters. Each team of four has a BA controller detailed.

Fig 4-4. Firefighting Organization (State 1) (Emergency Stations are Similar) - DD/FF and Smaller ships

CHAPTER 5

FIREFIGHTING EQUIPMENT

0501. 9 Litre Stored Pressure Extinguisher (AFFF) (SPE(AFFF))

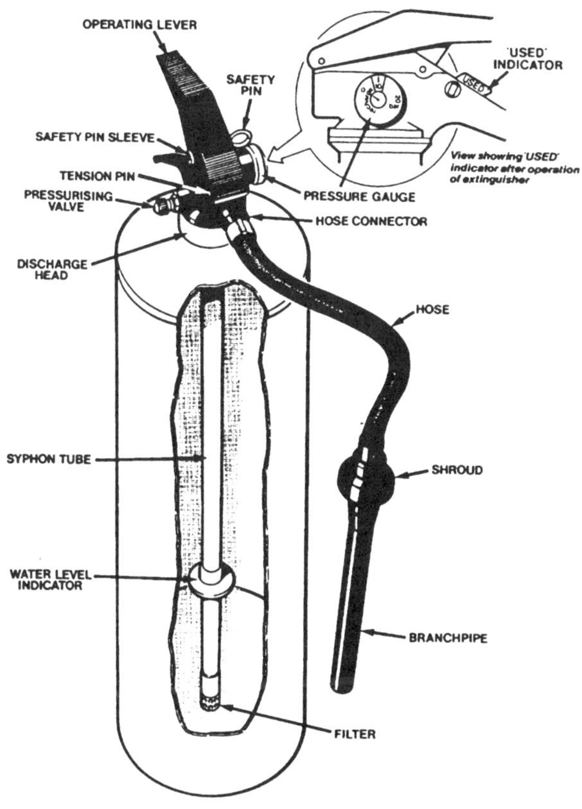

Fig 5-1. 9 Litre Stored Pressure Extinguisher (AFFF) - SPE(AFFF)

CAUTION
THE INSTRUCTIONS ON THE SPE(AFFF) ARE IN RED, THOSE ON THE SPE(DP) ARE IN BLUE.

Firefighting equipment

This appliance is used for fighting fires involving burning solids or liquids.

> WARNING
>
> **WHEN FIGHTING A FIRE IN THE VICINITY OF ELECTRICAL POWER THAT HAS NOT BEEN SWITCHED OFF ENSURE THE NOZZLE OF THE SPE(AFFF) IS NOT NEARER THAN 1.8m TO THE LIVE EQUIPMENT.**

The discharge from this appliance can be controlled by means of the operating lever which can be released whilst it is being re-directed, thus preventing wastage. The appliance is refilled as follows and recharged as in paragraph 0502.

a. Ensure that all pressure has been released and unscrew the discharge head assembly, complete with the syphon tube, and withdraw it from the shell.

b. Wash out the inside of the shell thoroughly with fresh water.

c. Put 8.6 litres of clean, fresh water into the extinguisher using the syphon tube as a dipstick. The correct amount of water is indicated when it reaches the end of the syphon tube when the measuring shoulder, which projects out from the side of the syphon tube, is rested on the lip at the top of the extinguisher. Using a AFFF measuring cylinder (graduated, acrylic, 500 ml - from the fire locker) top up the extinguisher with 0.4 litres of AFFF thus making the contents 9 litres.

Note. *Should sea water ever be used to fill the extinguisher in an emergency, it is to be well rinsed out with clean fresh water after use, and before refilling and recharging.*

d. When the AFFF solution level is correct, replace the discharge head assembly and hand tighten.

Note. *A thin film of grease (Vaseline) is to be applied to the head seating 'O' ring before replacing the head assembly.*

e. Raise the operating lever to its non-operational position, insert the safety pin and retain it by fitting a Hellerman rubber sleeve to the end of the pin.

f. Re-set the USED indicator.

g. Remove the air charging valve dust cap and connect the compressed air supply to the charging valve.

h. After charging (see paras 0502, 0503, or 0504) disconnect the air supply and replace the dust cap. Release excess air as in the note with para 0504.

i. With the SPE in the upright position check for signs of leakage around the cylinder head.

j. With the nozzle stow and secure the SPE, using the arrangements provided.

0502. Pressurization Arrangements - SPE(AFFF)

Pressurization is normally carried out using the portable air charging assembly, the ship's breathing air charging panel or, in an emergency, a foot pump. These systems are described below.

0503. SPE(AFFF)'s are to be charged to 10 bar - the working pressure - Indicated within the black segment on the extinguisher pressure gauge. The performance of the SPE(AFFF) at 7 bar is acceptable in an emergency. This is achievable using the tyre pressure gauge and the foot pump.

0504. Portable Air Charging Assembly

This assembly (see Fig 5-2) consists of a 1240 litre free air capacity steel high pressure air cylinder fitted with a regulator valve, pressure gauges and a discharge hose terminating in a connection which will mate with the air charging connection on the SPE. The assembly is also fitted with a carrying handle and wind around hose stowage. The air cylinder is charged to 207 bar with clean, dry air from the ships breathing air charging panel, if fitted, or from the ships usual charging facilities ashore and will charge an SPE to 10 bar.

Note. In the event that an SPE is overcharged, especially when using this method, the excess pressure should be released by standing the SPE upright and depressing the valve stem inside the air charging connection (Schraeder valve).

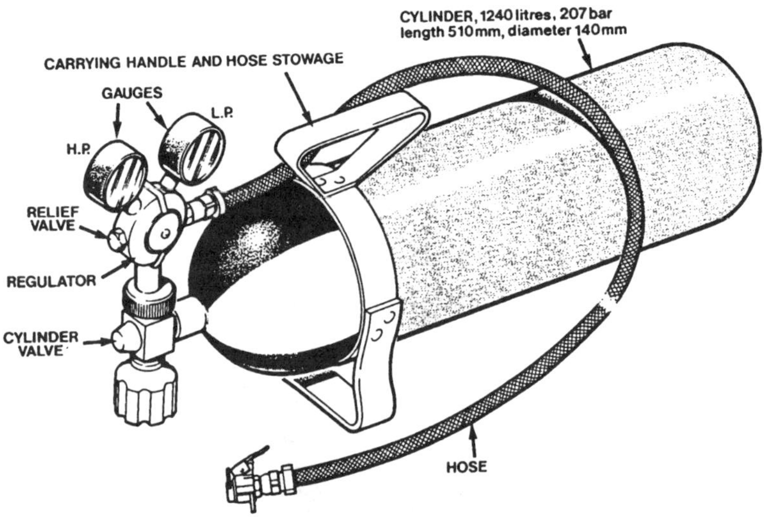

Fig 5-2. Portable Air Charging Equipment

0505. Foot Pump

In an emergency the SPE(AFFF) may be charged using a foot pump. The method of charging is self evident in that the pump discharge hose is connected direct to the SPE charging connection. The pressure to be achieved in the SPE is 7 bar.

0506. Stored Pressure extinguisher (Dry Powder) (SPE(DP))

This extinguisher is provided near to, or in, the accesses to some machinery spaces, in the flight deck area and in hangars for fighting liquid fuel fires. It is especially useful for fighting liquid fuelled pressure spray fires. The discharge from this appliance can be controlled by means of the operating lever which can be released whilst it is redirected, thus preventing wastage. It must be pressurized to 10 bar with DRY air from the portable charging assembly.

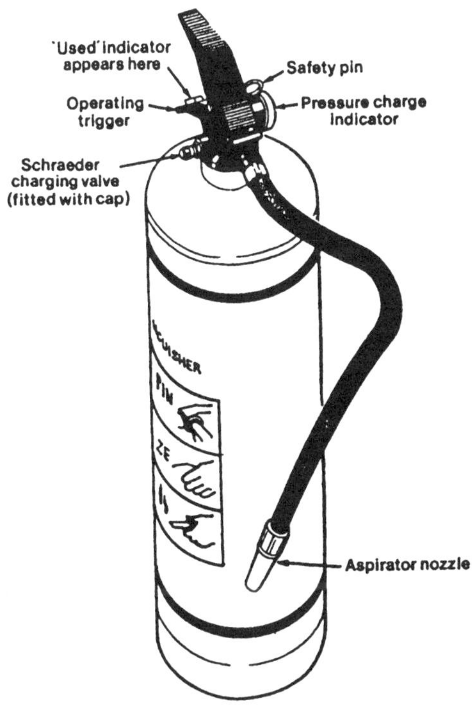

Fig 5-3. 9 Litre SPE (Dry Powder) - SPE(DP)

0507. SPE(DP) Filling

The SPE(DP) should be recharged with dry powder in a dry compartment as follows:

a. Ensure the extinguisher is full discharged.

b. Remove the discharge head assembly, clean of all powder residues with a soft brush.

c. Clean the Schraeder valve assembly, if necessary removing the valve core.

d. Clean out the cylinder using a soft brush ensuring the inside of the cylinder is dry. Particular attention is to be paid to the threads and joint sealing faces.

e. Using a suitable chute as a funnel pour the contents of one 25 lb bag of powder into the cylinder.

f. Check cleanliness and absence of powder from the screw threads and joint faces, then re-fit the head assembly.

g. Pressurise with DRY air, via the Schraeder valve, to 10 bar.

Note. Do not use a foot pump in this (the SPE(DP)) application.

0508. CO_2 Fire Extinguisher 2 kg (See para 0603a)

WARNING

WHEN OPERATING CO_2 FIRE EXTINGUISHERS HOLD ONLY BY THE OPERATING HANDLE AND ANY THERMALLY INSULATED GRIP ON THE NOZZLE. HOLDING THE CYLINDER OR THE HORN MAY WELL CAUSE ICE 'BURNS'.

a. CO_2 fire extinguishers provided in compartments containing high voltage electrical equipment are fitted with a flexible hose and plug, which fits an inlet socket on the electronic cabinets.

b. For use on open equipment and elsewhere in these compartments the hose is fitted with a hand applicator, which fits over the plug designed to prevent hand contact with the end of the hose (see Fig 5-4).

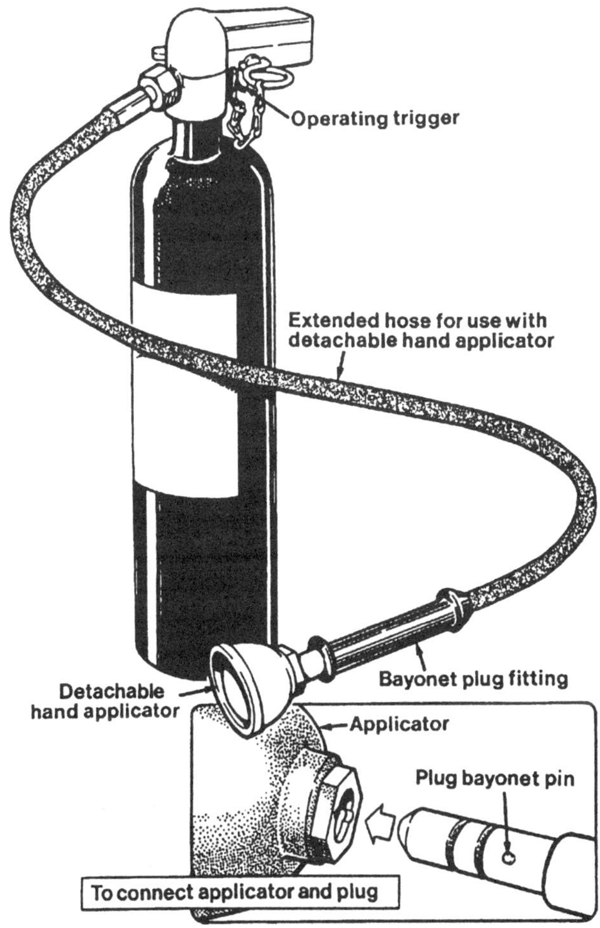

Fig 5-4. CO_2 Fire Extinguisher

c. Elsewhere in the ship CO_2 fire extinguishers fitted with a rigid horn are supplied for dealing with electrical fires (see Fig 5-5).

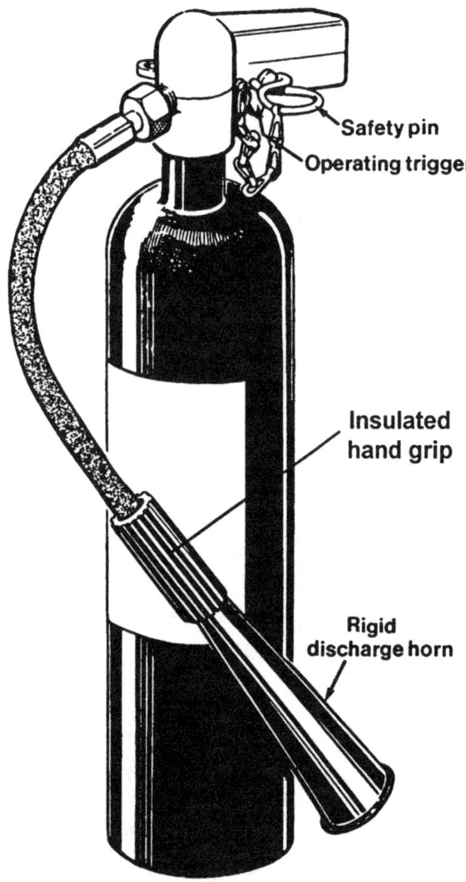

Fig 5-5. CO_2 Fire Extinguisher with Rigid Discharge Horn

d. CO_2 fire extinguishers cannot be recharged onboard. When discharged, or partially discharged, they are to be returned through naval stores for recharging.

0509. Spare

NOZZLES

0510. The types of nozzle provided are:

a. **Firefighter Nozzle.** This is provided at all fire hydrants, in all ships, except on weather decks. It is stowed in the shut position, that is, the rotating sleeve turned fully anti clockwise (looking from the operators end), bringing the rotating sleeve toward the operator. Opening the nozzle by rotating the sleeve clockwise one quarter of a turn produces a waterwall. A further quarter of a turn produces a ragged jet of water to fight the fire. This nozzle has a low water discharge rate and is used for fighting fires.

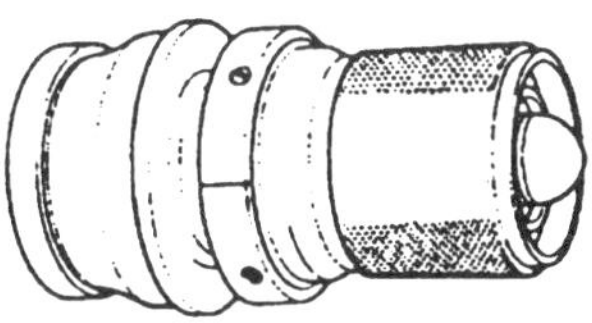

Fig 5-6. Firefighter Nozzle

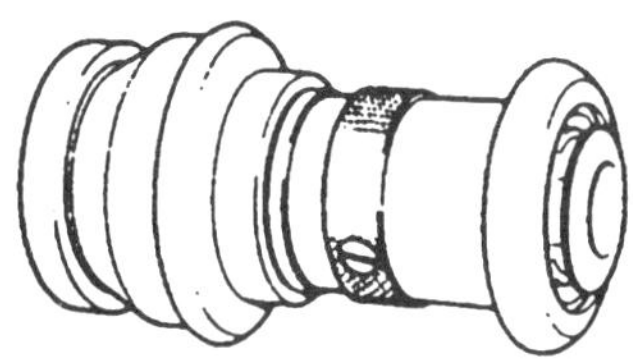

Fig 5-7. Waterwall Nozzle

b. **Waterwall Nozzle.** This is provided at all weatherdeck hydrants and three are kept in each FRP fire locker for use in providing waterwalls. It is stowed in the shut position which is achieved by turning the rotating sleeve clockwise (looking from the operator) which brings the rotating sleeve towards the operator. Opening the nozzle by rotating the sleeve anti clockwise produces, initially, a jet, but the nozzle must be opened fully to produce the desired waterwall.

c. **Fixed Hatch Waterwall Nozzle**. Fixed hatch waterwall nozzles are fitted in the corners (their spray angle is approximately 90°) of all hatches leading to high fire risk compartments to facilitate re-entry when firefighting by providing a more effective sealing waterwall thus reducing the number of personnel in close proximity to the hatch.

0511. Hose Couplings

Firefighting hoses are joined together and connected to the HPSW main by instantaneous couplings. Each hose has a male coupling at one end and a female at the other. All hydrants have female connections so that, when coupling up to a hose the male end of the hose must be taken to the hydrant and the female end towards the fire.

0512. Stowage and Care of Hoses

A hose should, at all times, be coupled to each hydrant and a (shut) nozzle connected so that water can be delivered immediately when it is required. Although hoses are fairly robust they need care if they are to be kept in good order, and the following instructions must be observed:

a. Hoses are to be faked out in their stowages ready for running out (see Fig 5-8). Where double hose baskets are fitted the hose in the section of the basket closest to the bulkhead must be faked and the hose in the basket furthest from the bulkhead rolled.

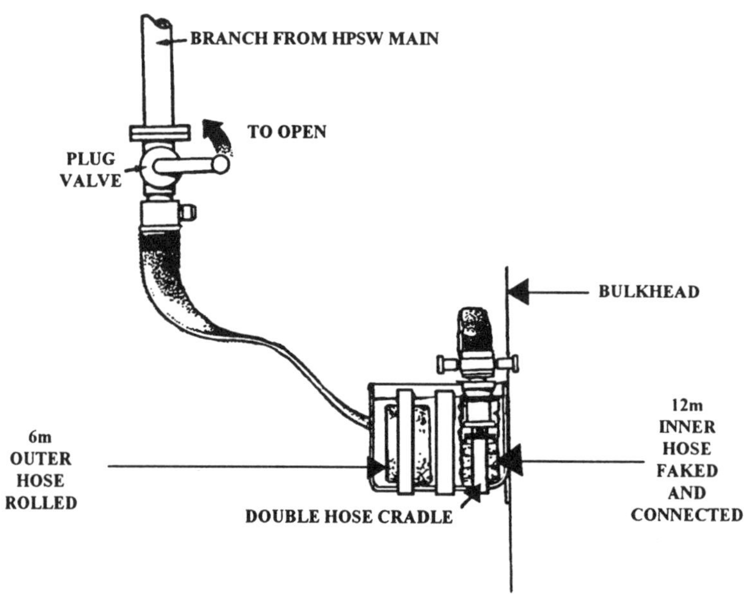

Fig 5-8. Stowage of Hoses

b. Charged hoses should not be dragged over rough surfaces, especially when kinked.

c. Hoses may chafe (due to vibration) and kink where they first touch the deck after leaving a portable pump and will also tend to kink when leaving hydrants. Arrangements should be made to prevent kinking, which reduces water flow, and to protect hoses from chafing by wrapping rags round the hose at the deck contact point. Similarly, protective padding should be used where hoses pass over sharp edges such as watertight door of hatch coamings.

d. To avoid subjecting hoses to sudden shock or strain, hydrants and pump delivery valves should always be opened slowly. The sudden closure of nozzles should be avoided.

e. Hoses must be drained and wiped down before re-stowing. To expel water the hoses must be under-run at shoulder height. When re-stowing the position of the previous folds should be shifted and the nozzles turned to the shut position (see para 0510).

f. Hoses should be washed and flushed after contamination with fuels, oils, greases, and after use with AFFF.

0513. Foam Producing Appliances

These consist of foam making branchpipes, the FB5(X) Mks 1 and 2, the FB10(X) and the FB10/10 (a commercial replacement for the FB10(X)), the incline inductors FBU(X)5 and FBU10(X) (fitted only in or near hangars) and the centre feed hose-reel (CFHR), details of which are as follows.

a. The **FB5(X)** is a lightweight foam making branchpipe for general use against liquid fuel fires (except in the hangar or on the flightdeck where the larger FB10(X)/FB10/10 is used). The pick up assembly (suction hose and spill) is connected by a quick release coupling. To induce AFFF the cap must be removed from the 20 litre container and the spill inserted. This drill is repeated using back up containers as the previous container empties. This branch pipe will also put foam into machinery spaces via the foam inlet tubes. The FB5(X) should only be used with No 2 hoses (No 1 hoses restrict the flow rate and are susceptible to kinking). One drum of foam lasts approximately 1 minute.

b. The **FB5(X) Mk 2** has an on/off facility which allows the hose to be fully charged but the flow shut off at the nozzle. A non return valve in the pick up spill prevents water flowing back into the AFFF drum. The Mk 2 is also fitted with a foam mix selector which, for RN use, must be pinned at the 6% AFFF position. This is achieved when the arrow points to the 6.

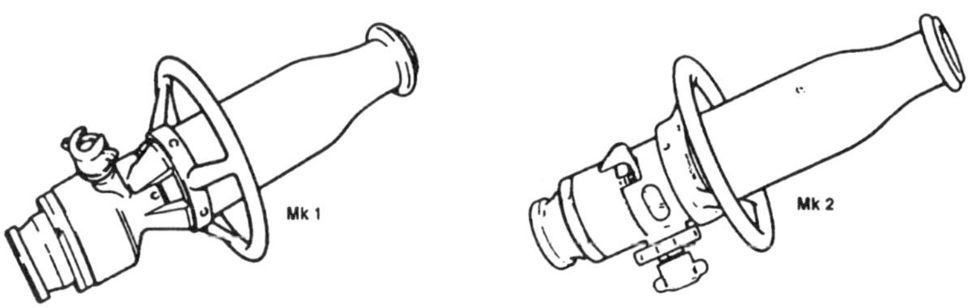

Fig 5-9. Foam Making Branchpipes FB5(X) Mks 1 and 2

c. The **FB(10)X foam making branchpipe** (see Fig 5-10) and its successor the **FB10/10** (which is a commercial variant, similar in appearance and performance to the FB10(X)) are high output branchpipes used at helicopter landing positions, on flightdecks and on vehicle decks. Hydrants at these positions have an inline inductor (the FBU10(X)) bolted to the flange before the instantaneous coupling. The pick-up assembly is attached to the inductor and has a shut off cock. The inductor passes the mixture of AFFF and water through the hose to the branchpipe which aerates and delivers the foam. To produce foam of good quality the FB(10)X needs HPSW mains pressure of at least 4.2 bar. This branchpipe can deliver foam to a distance in the form of a jet, or less forcibly as a spray.

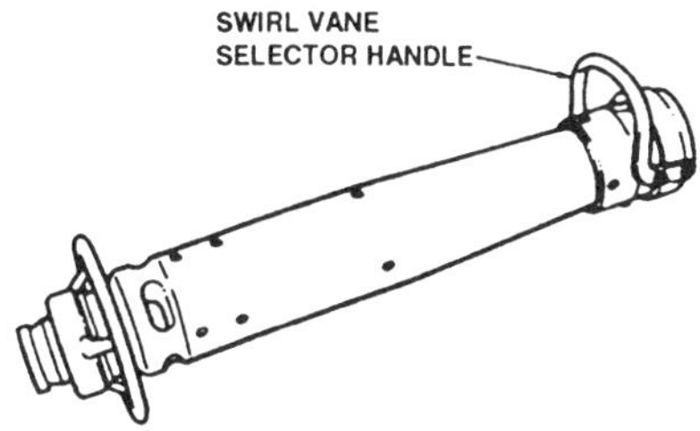

Fig 5-10. FB(10)X Foam Making Branchpipe

0514. Portable Inline Inductor (FBU5(X))

This inductor (see Fig 5-11) is designed for use with the FB5(X). Its purpose is to allow AFFF to be introduced into the hoseline at, or near, the hydrant position and not directly into the branchpipe. This allows the operator much greater mobility in that he is no longer constrained by the AFFF containers and the pick up assembly, especially when passing through WT openings or using ladders. The inline inductor has a stowage bracket with a 1.8m length of No 2 hose (for use between the hydrant and the inductor). The pickup spill for the inductor is taken from the FB5(X), which no longer needs it when being used in conjunction with the inline inductor. The spill connecting point on the FB5(X) need not be blanked off. The inline inductor should not be used when the FB5(X) is being used to introduce foam via foam inlet tubes. They are best kept for use with firefighting hoses which are being moved around by operators, as opposed to static hoses, eg those led to foam inlet tubes. A direction of flow indicator should be marked on the FBU 5(X) to avoid it being connected the wrong way round.

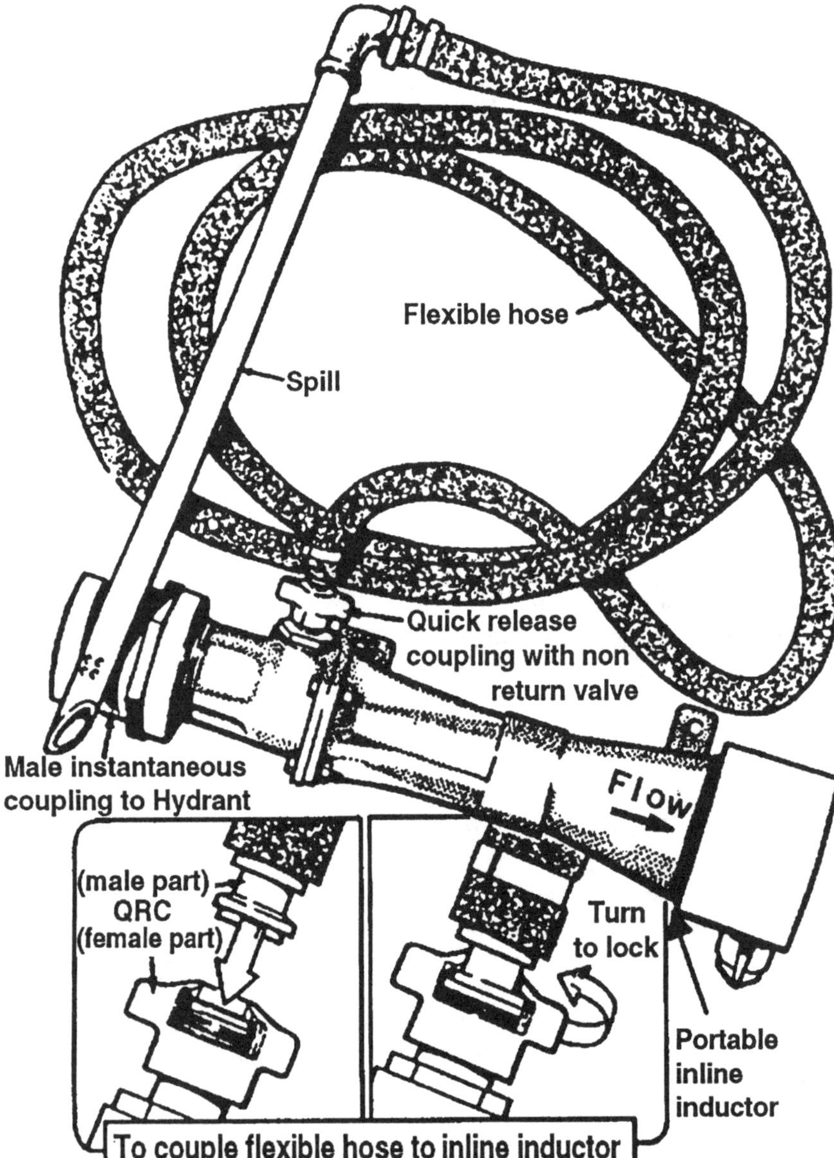

Fig 5-11. Portable Inline Inductor - (FBU5(X))

0515. Centre Feed Hose-Reel (CFHR)

This is a first aid firefighting appliance, normally used in the FOAM mode, but with the facility to be used in the WATER mode for boundary cooling. It is normally left in the STANDBY condition with:

a. Remote high pressure salt water isolating valves (if fitted) pinned or wired in the OPEN position.

b. Local/adjacent HSWP isolating valves may be closed.

c. The 3 position selector valve on the inductor set to the FOAM position.

d. The nozzle valve on the hose end set to the OFF position.

Notes:

1. A copy of the operating instructions and a deck plan of compartments served should be posted adjacent to every hose-reel assembly.

2. The FOAM (F) and WATER (W) positions of the valve and the outboard end of the operating handle are to be marked with photoluminescent paint.

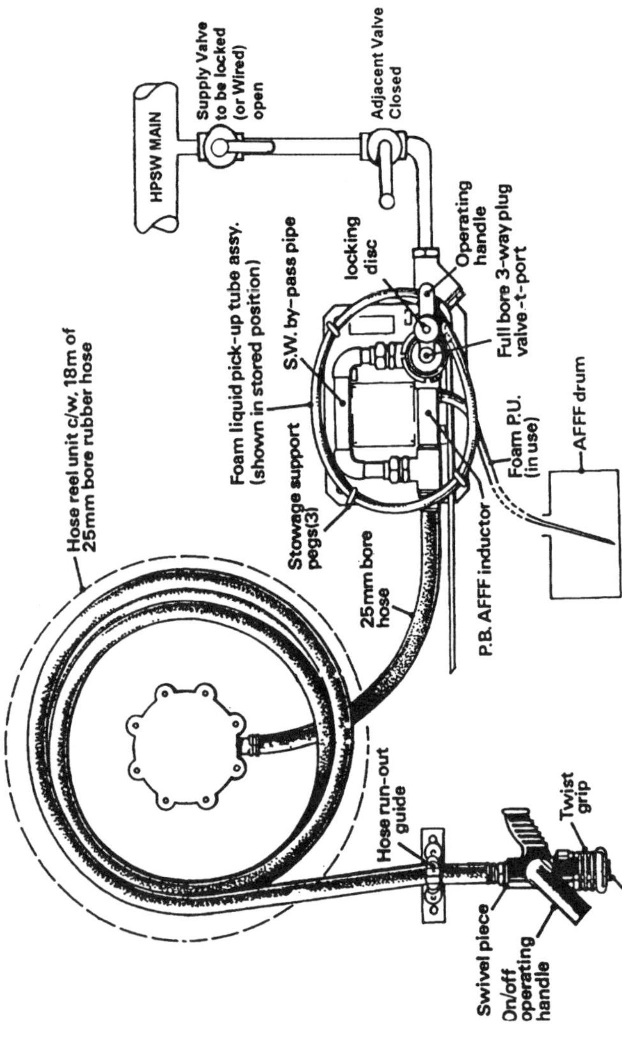

Fig 5-12. Centre Feed (Foam/Water) Hose-Reel Unit

CHAPTER 6

FIREFIGHTING TACTICS AND TECHNIQUES

At all times personnel and equipment must be organized so that the attack on a fire will be immediate, continuous and aggressive.

USE OF FIRST AID FIREFIGHTING APPLIANCES

0601. Solid Fires

Solid fires are to be attacked using either the centre feed hose-reel (CFHR) using its AFFF inductor or a 9 litre stored pressure extinguisher (AFFF), (SPE(AFFF)). These appliances are to be used as follows:

a. **Centre Feed Hose Reel Operating Instructions**

(1) *First Aid Firefighters.* The FOAM mode is to be used for all first aid firefighting operations with this equipment.

(a) Remove the cap from the AFFF drum and insert the spill pick up.

(b) Open the isolating valve adjacent to the inductor panel.

(c) Proceed to the location of the fire.

(d) Direct the nozzle away from the fire and open the nozzle valve - in jet mode to produce foam.

(e) When foam is being produced attack the fire - changing to ragged spray/cone effect.

(f) The next man on the scene must re-supply AFFF.

(2) *Boundary Cooling.* The WATER mode is to be used when boundary cooling.

(a) Select WATER position on the selector valve.

(b) Proceed to the location requiring boundary cooling.

(c) Open the nozzle and select ragged spray.

(d) Boundary cool using the minimum amount of water.

b. **9 Litre SPE(AFFF).** Remove from its stowage and check that the gauge indicates an adequate charge. Holding the appliance in front of the body, approach the fire keeping as low as possible. When in a safe discharging position (with a clear withdrawal route), and as close to the fire as possible, free the hose, pull out the safety pin and direct the nozzle over the fire beating the flames down. Then direct the discharge to the heart of the flames, breaking open the combustibles as much as possible. A SPE charged to a pressure of 10 bar will give a maximum throw of 7.5m and last for approximately 50 seconds. An SPE charged to 7 bar will give a maximum throw of 6.5m and lasts approximately 40 seconds.

0602. Oil Fires

Oil fires are attacked, in the first instance, as described in para 0601, except where dry powder extinguishers are more appropriate and available (some machinery space accesses and on flightdecks and in hangars) especially on liquid fuel fires which are under pressure and producing a spray. The 9 kg Stored Pressure Extinguisher (Dry Powder) (SPE(DP)) is similar in appearance and operation to the SPE(AFFF), the difference being that its instruction legend is in blue (instead of red) and with 1" wide blue bands around the extinguisher body, above and below the legend. It is operated in the same manner as the SPE(AFFF) by removing the safety pin and checking that ten bar is registered on the gauge, press the trigger and employ a sweeping horizontal motion with the nozzle directed at the base of the fire. **A continuous discharge lasts for approximately 20 seconds.**

0603. Electrical Fires

Wherever possible the equipment and/or the circuits on fire should be isolated. Delays in achieving this should not prevent the fire from being attacked in the following manner:

a. Electrical and electronic equipment fires should be attacked using 2 kg CO_2 fire extinguishers. The extinguisher is operated holding it by the squeeze grip release valve, removing the safety pin and then pressing the trigger of the grip very firmly to ensure full discharge and to avoid 'freezing up' at the orifice. The CO_2 discharge should be directed at the base of the fire from as near as possible, ensuring that the insulated handle on the discharge horn is used as the grip.

b. Other electrical fires can be attacked using AFFF first aid appliances (SPE or CFHR) before the equipment or circuits have been isolated providing the voltage does not exceed 440V ac or 800V dc, and the appliance is at least 1.8m from the power source.

Note. *In the event of minor fire symptoms being noticed, eg, wisps of smoke coming from electronic equipment, switchboards, etc, switching of the power supply may be sufficient to cause these symptoms to disappear. Discretion is therefore necessary in initial firefighting action. Wherever possible electrical fires should be fought using CO_2 gas.*

0604. Ventilation Control

In the event of fire the local starters to ventilation systems, which are not controlled by breakers, should be tripped. These systems may be re-started on the authority of HQ1. Ventilation controlled by switchboards/SCC/HQ1 is crash stopped and then controlled in accordance with ship's orders.

Note. *ACU's fitted to individual compartments (and for that compartment only) and full recirculation ATU systems for unmanned spaces, should not be stopped.*

0605. Smoke Control

A smoke boundary can be set up by closing appropriate doors, hatches, ventilation flaps and valves and, most importantly, the fitted smoke curtains which are also used to provide access and exit from the fire zone. Remember, under some circumstances the ventilation systems can be used to remove or contain smoke.

0606. Fire Boundary Control

All flammable material and dangerous items, eg, compressed gas bottles, chemicals, etc, should be removed from the compartments adjoining the fire boundary to safer areas.

0607. Fire Boundary Cooling

Once the fire boundary has been established, arrangements should be made to prevent the conduction of heat via bulkheads, decks and trunkings and so contain the spread of the fire. This is achieved as follows:

a. Bulkhead and deckhead linings on the fire boundary should be removed, or access gained through them, in order to allow the cooling spray to reach the heated structure.

b. Apply the minimum amount of cooling water, with the FF/CFHR Nozzle, sufficient to keep the plating moist.

c. Never leave cooling appliances unattended.

d. Prevent the accumulation of free surface water. It should be pumped overboard or drained to the lowest, smallest, watertight compartment possible.

e. To prevent boundary cooling compartments becoming untenable, use exhaust ventilation (with HQ1/SCC permission) to remove steam.

f. Cover electrical equipment on the cooling boundary, using plastic sheet (polyamide, Nylon 6, transparent protective film) to protect it from water. Switch off high voltage electrical equipment sited on the cooling boundary (with HQ1/SCC) permission.

g. Monitor bulkhead temperatures using "Rototherms". Prior to making a re-entry confirm:

(1) Three successive temperature drops have occurred, and/or

(2) Compartment temperature is below 100 degrees Celsius.

0608. Fire Approach Technique

Sometimes the position of the fire and the layout of the ship dictate that only one method of approach is practicable on order to attack it. Whenever there is a choice, the methods preferred are in the following order:

a. **Approach From the Same Level.** This is the preferred route to any fire as the firefighting equipment is easier to handle and access/exit is more convenient.

(1) Solid fuel fire - see Fig 6-1.

(2) Oil fuel fire (diesel, hydraulic oil, etc) - see Fig 6-3 (which illustrates a four man team, with an FB5(X) instead of the Firefighter nozzle, and an extra man, No 5, on the waterwall), and adapt this drill to suit an entry via a door (see Fig 6-1).

b. **Approach from Above.** The firefighters will encounter a smoke/heat layer which deters all but the most experienced and determined firefighters. Invariably access to the fire is gained via a hatch as follows:

(1) Solid fuel fire - see Fig 6-2.

(2) Oil fuel fire (diesel, hydraulic oil, etc) - see Fig 6-3.

c. **Approach from Below.** This is the coolest and most smoke free route to a fire but is rarely available on a ship. Difficulties will be experienced opening hatches above in order to access the fire area. It is an energy consuming and difficult approach which is compensated for by lack of smoke, lower temperature and the fact that, once the hatch is opened, water or foam can be bounced off the deckhead or bulkheads towards the seat of the fire.

0609. Premature Withdrawal of Firefighters

Should the warning whistle on the BA of a member of the firefighting team sound:

a. Prematurely, namely before the pressure gauge indicates 7 minutes of air remaining in the set; or

b. When 7 minutes of air are indicated in the set and the relief firefighting team has not arrived; or

c. One or more of the firefighting team has to withdraw due to injury or some unforeseen circumstances.

then this is to be communicated to the FRP (allowing whoever is in charge of the firefighting teams to make arrangements for timely relief) and the entire team must withdraw ensuring minimal support for the fire, eg closing doors and hatches behind them. The relief team will then recommence the attack of the fire using the appropriate procedures for opening doors and hatches. **Wherever possible the attack on the fire must be maintained continuously and aggressively.**

0610. Fire Sentries

When a fire has occurred and has been extinguished an appropriate number of suitably brief and equipped sentries must be posted in order to deal with, and raise the alarm, in the event of re-ignition.

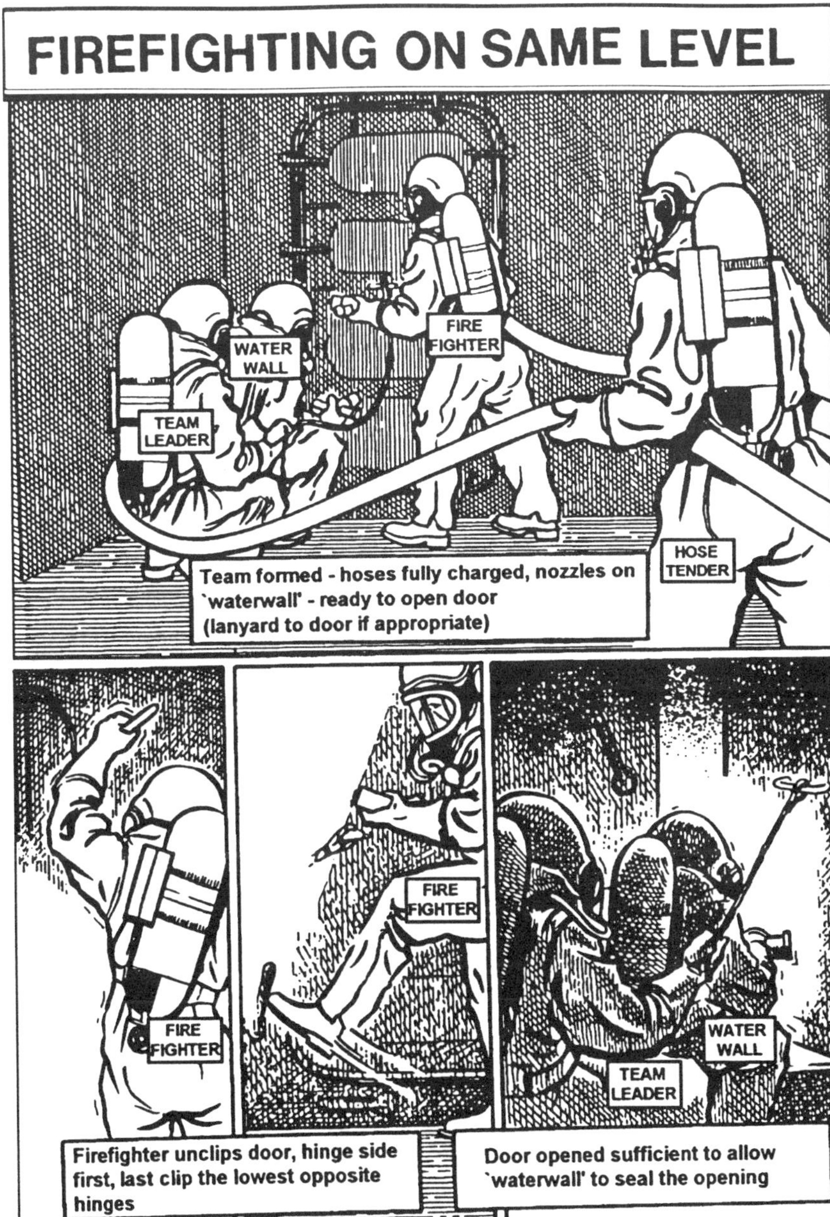

Fig 6-1a. Firefighting on The Same Level

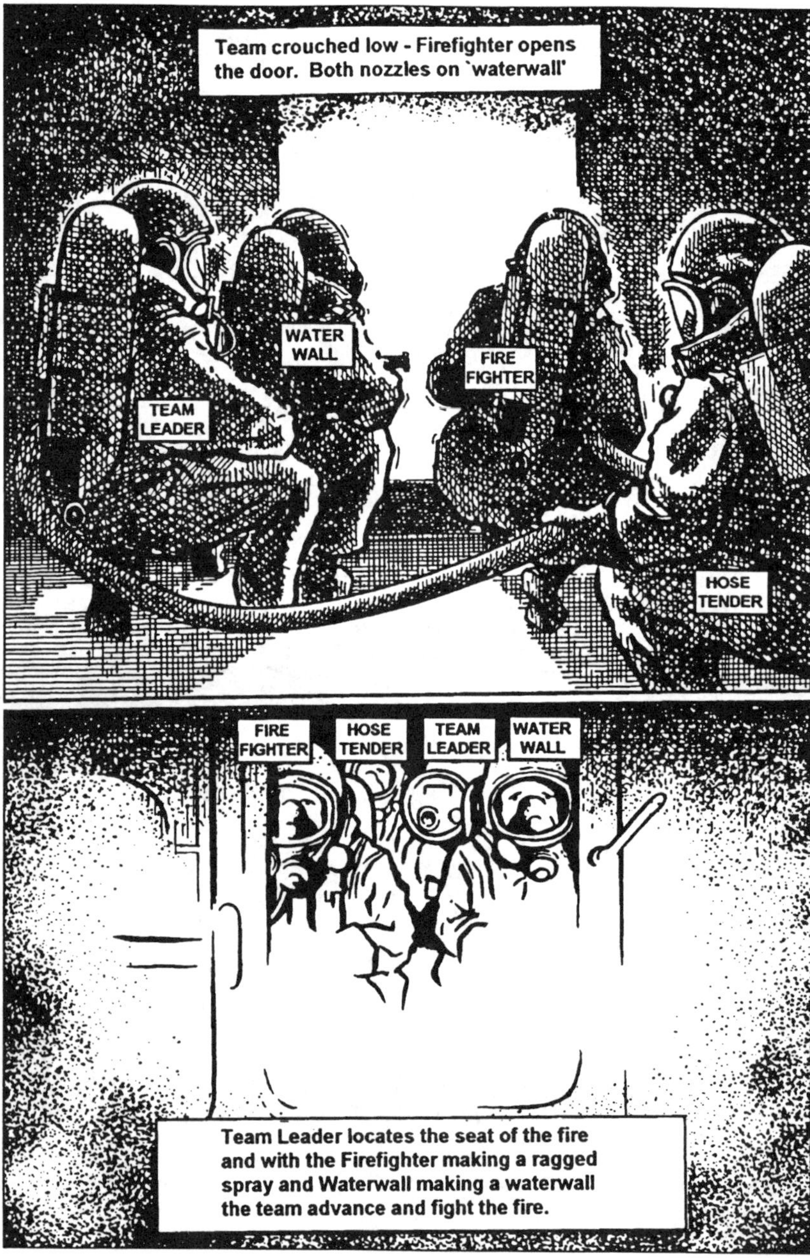

Fig 6-1b. Firefighting on The Same Level

Fig 6-2a. Firefighting From Above

Fig 6-2b. Firefighting From Above

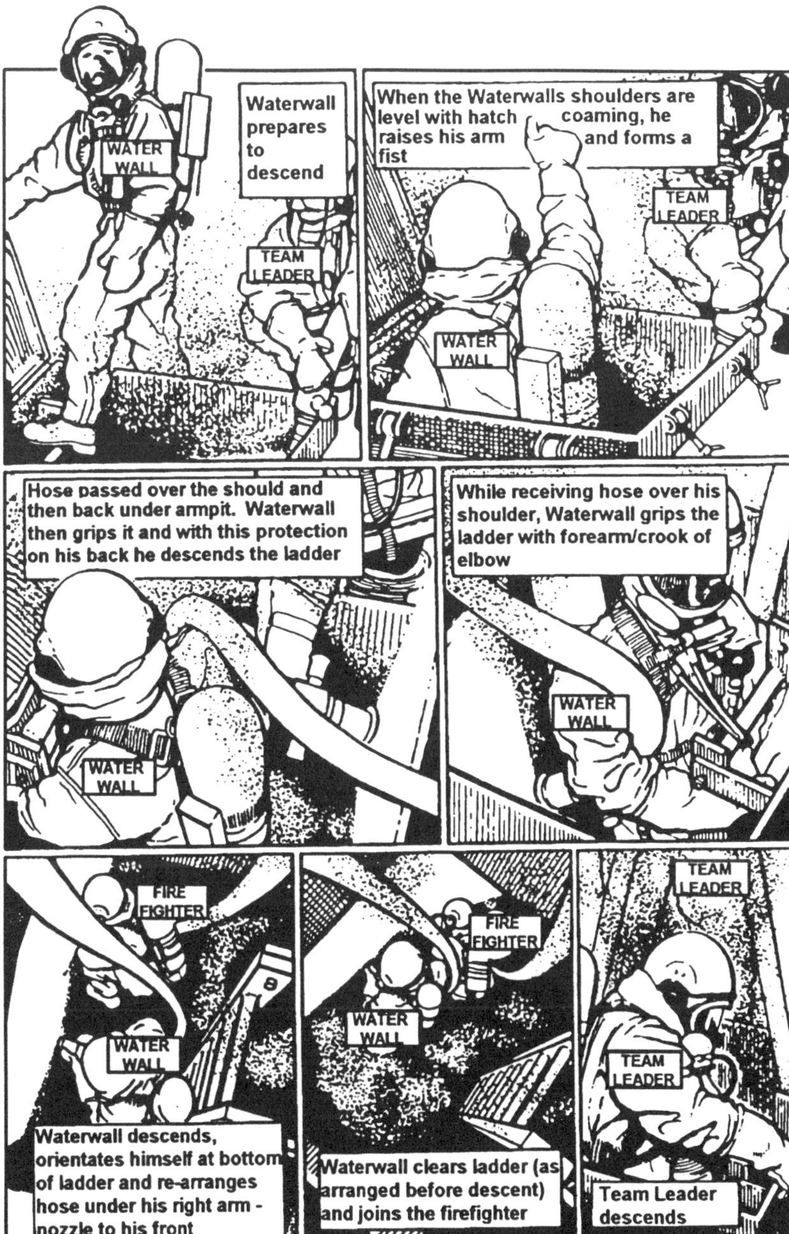

Fig 6-2c. Firefighting From Above

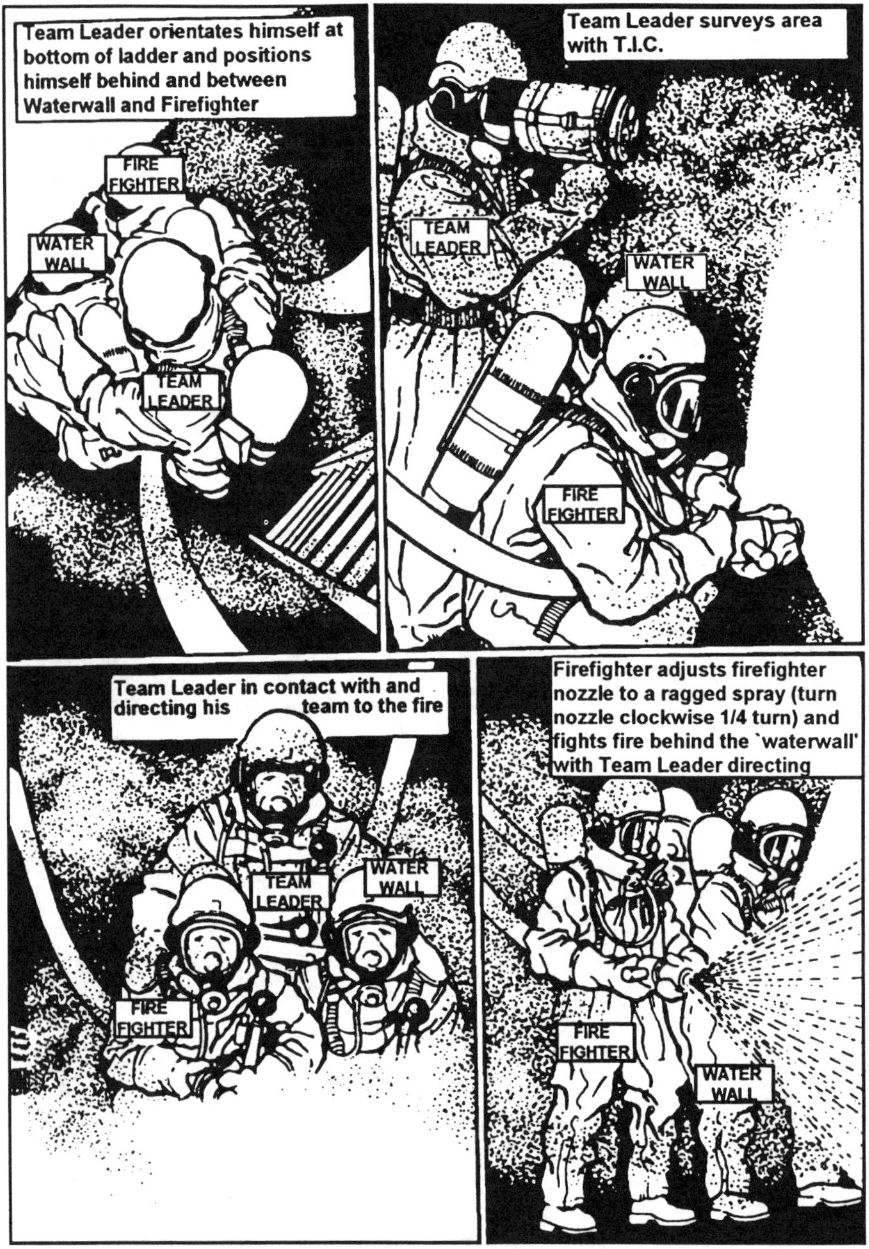

Fig 6-2d. Firefighting From Above

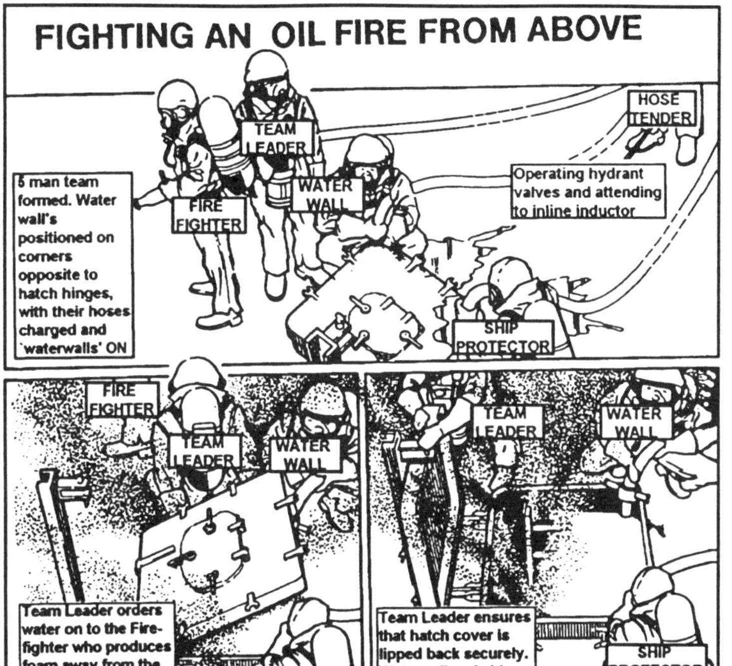

The routine shown is with the Mk 1 FB5X, if the Mk II is used the Firefighter is able to shut off his FB5X prior to entry and re-open it once directed to by the Team Leader

Fig6-3a. Firefighting an Oil Fire from Above

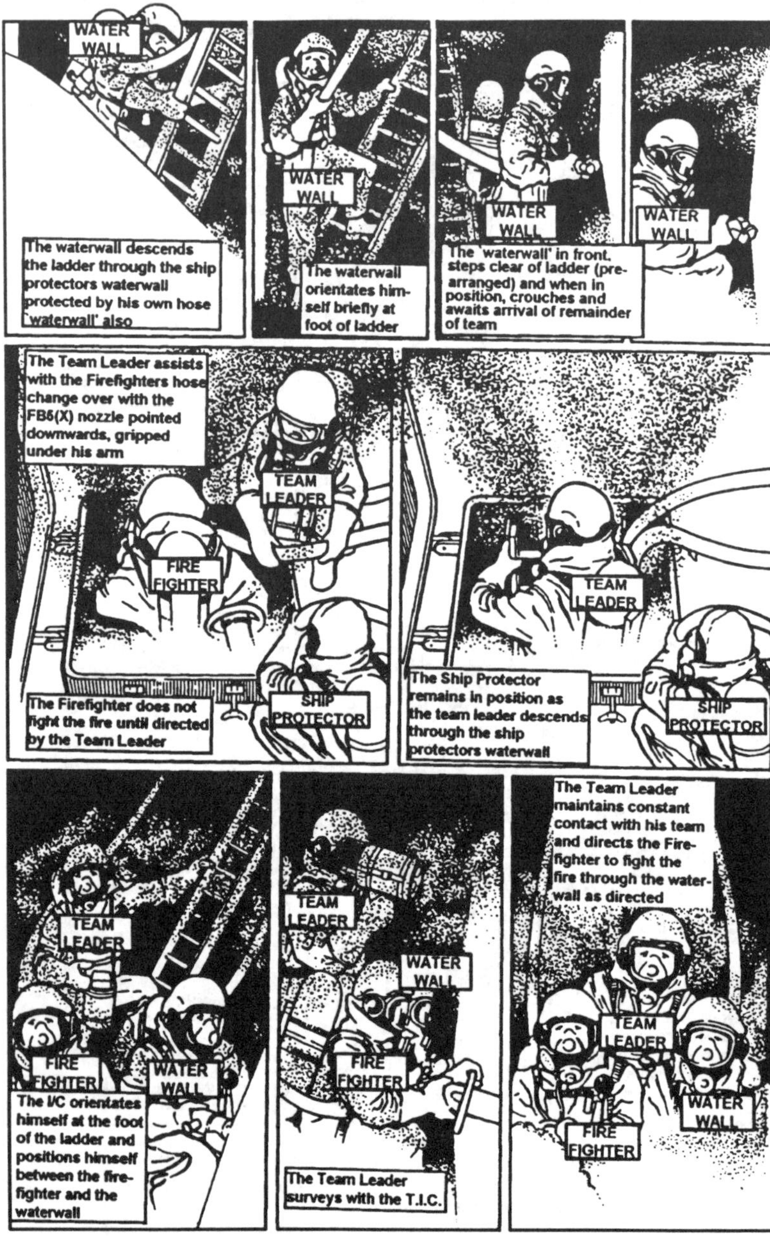

Fig 6-3b. Firefighting an Oil Fire from Above

CHAPTER 7

PERSONAL PROTECTION

0701. Introduction

Firefighting personnel need to be protected, as far as is possible, against the effects of:

- heat;
- toxic gases or lack of oxygen;
- loss of vision;
- loss of communication;
- physical damage, especially to the head.

This is achieved by wearing the protective clothing and equipment that comprises the full firefighting rig.

PROTECTIVE CLOTHING - GENERAL

0702. Basic Firefighting Rig

This consists of Proban treated Cotton No 8's, or cotton overall's, DMS boots, cotton underwear and woollen socks. When engaged in firefighting operations collars should be buttoned up and the sleeves rolled down and cuffs buttoned. **Nylon clothing must not be worn by firefighters.**

0703. Full Firefighting Rig

This consists of basic firefighting rig plus anti-flash hood and gloves, Fearnought suit and firefighters gloves, seamen's stockings, firefighting boots, BA and bulb horn, slim tank helmet (with caplamp) and portable radio transceiver. (See Figs 7-1 and 7-2.)

Notes:

1. When wearing the above remember to occasionally shrug within the clothing and move the tight spots. This will allow the air enclosed to circulate more freely and improve comfort.

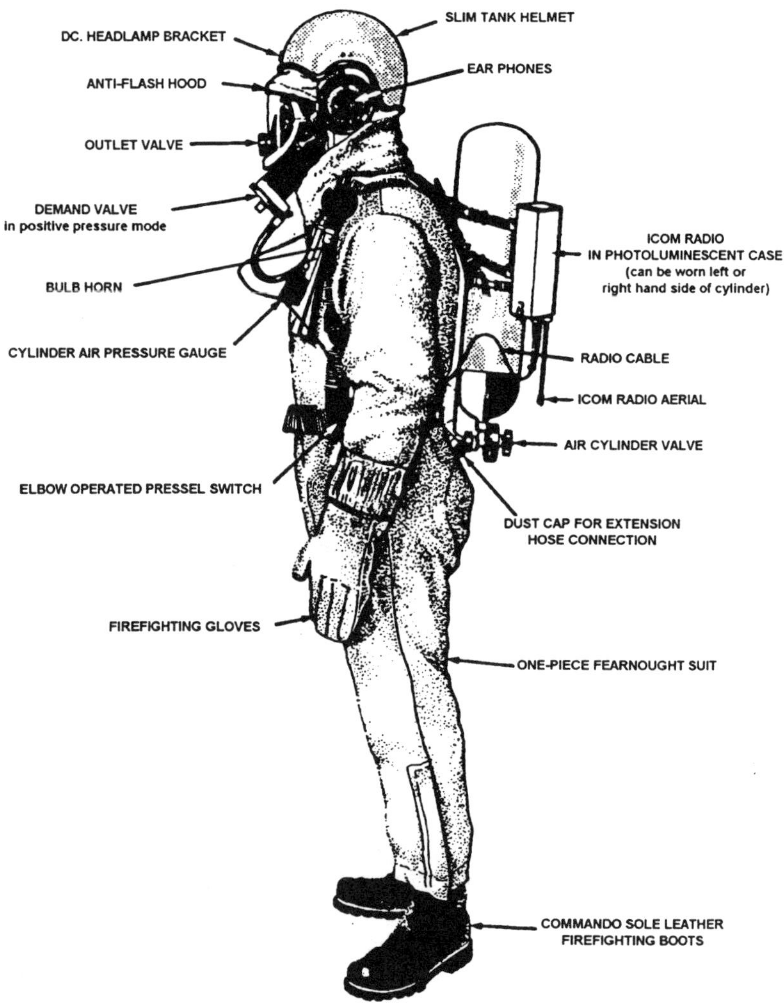

Fig 7-1. Full Firefighting Rig (Left View)

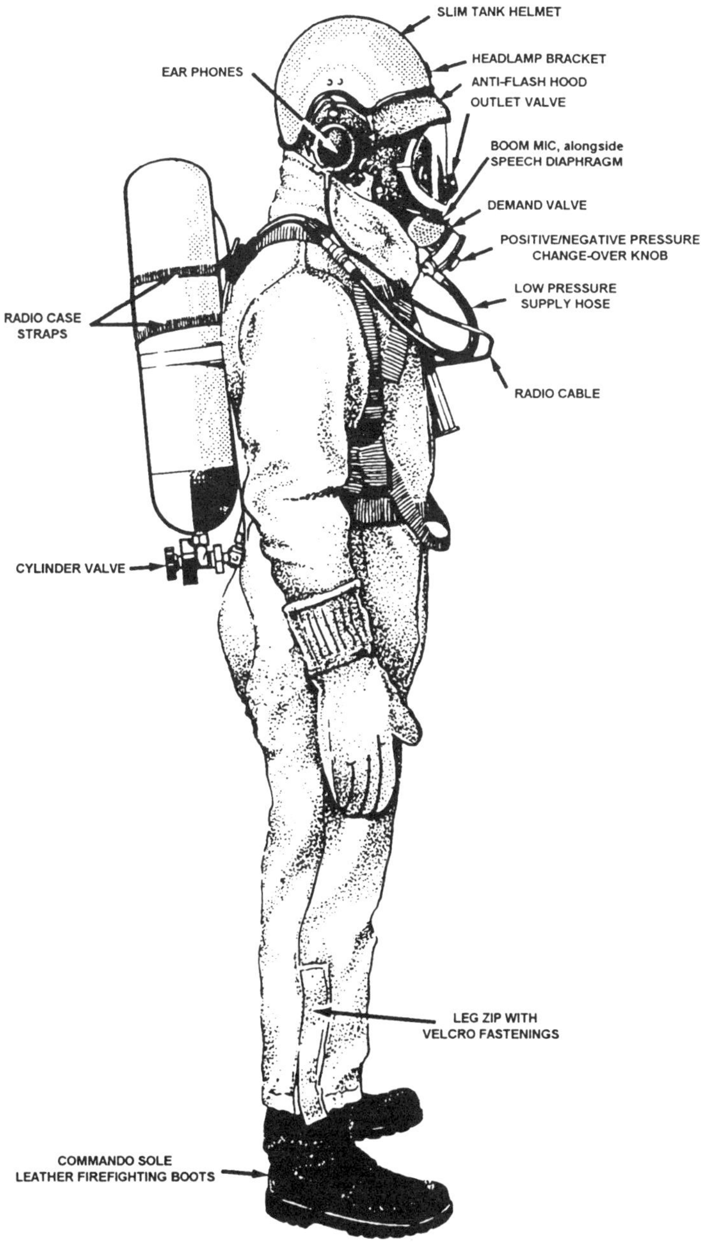

Fig 7-2. Full firefighting Rig (Right View)

2. Remember that when your ears start to tingle it is a warning of rise in body temperature and that steps should be taken to vacate the area quickly. Too long an exposure to a great heat will induce heat exhaustion and sudden collapse.

3. If a fearnought suit becomes wet the heat will cause steam to be produced and the wearer will risk injury from scalding. Once a suit has been wetted it must be kept moist (to protect the wearer).

4. All items of clothing and equipment needed to complete the full firefighting rig are stowed in the FRP locker, BA locker and nearby.

FULL FIREFIGHTING RIG

0704. Clothing (see Figs 7-1 and 7-2)

a. **Firefighting Boots.** The firefighting boots are leather high leg boots, with a thick commando style sole, re-enforced toe cap and heel with two loops at the top making them easier to pull on.

b. **Anti-flash Hood and Gloves.** Ready use anti-flash hood and gloves are stowed in each BASCCA locker for use by the Attack Party. The hood is worn and its mantle tucked inside the collar of the shirt or overalls. When wearing BA the chin bag of the hood is to be tucked over the equalization valve below the chin of the facemask. Anti-flash must be kept clean and white to provide the designed level of protection. It must not be marked in any way eg felt tipped pen or paint.

c. **One Piece Fearnought Coveralls (Zirpo Treated).** These incorporate a zip fastener from the left knee to the right shoulder and another zip at the bottom of each leg, to facilitiate the dressing procedure.

d. **Firefighting Gloves.** These are worn over the top of the cuff of the fearnought suit.

0705. Breathing Apparatus - General (see Figs 7-1 and 7-2)

a. Breathing apparatus, Self Contained, Compressed Air (Damage Control and Firefighting) (BASCCA(DC&FF) are stowed in lockers which are painted red with BASCCA(DC&FF) in white lettering. These are sited near FRPP's, upper deck fire re-entry posts and in pairs in lobbies and airlocks of the weather decks. A non-magnetic version of BASCCA(DC&FF) is supplied in some MCMV's.

b. The safety control equipment (BA control board, personal line, BA control armband and bulbhorn) is to be stowed in the BA locker. Guidelines are stowed in the four BA lockers adjacent to the FRPP and also in the upper deck re-entry locker. These lockers are to be marked to indicate that guidlines are stowed within.

c. The use of equipments in a search and rescue (SAR) is detailed in BR 2170(1).

d. Emergency Life Support Apparatus (ELSA) is to be used only for escape from, or through, smoke logged compartments. ELSA is not to be used for firefighting or damage control work.

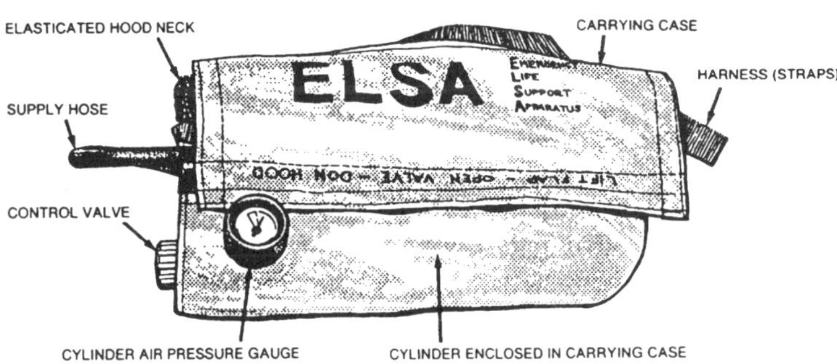

Fig 7-3. Emergency Life Support Apparatus (ELSA)

0706. Thermal Imaging Camera (TIC) - General

TIC's are used by the leaders of the four man firefighting teams. However, if they were required, they may previously have been used by the Attack Party.

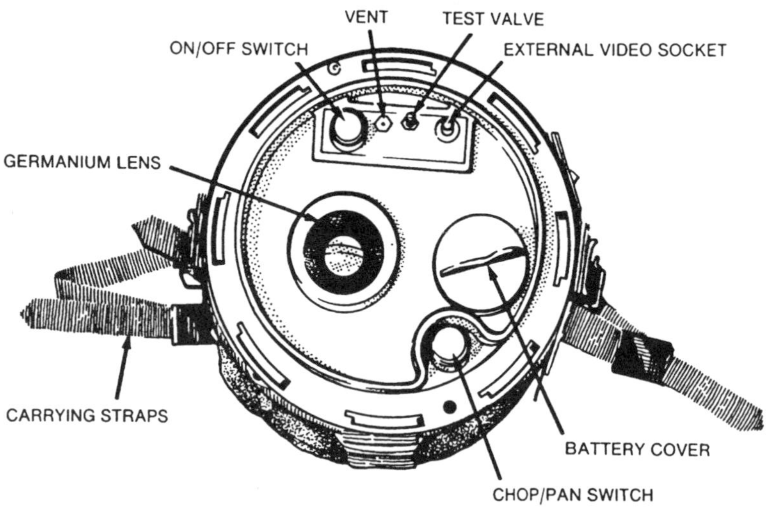

Fig 7-4. Thermal Imaging Camera

a. TIC's (large or miniaturized) are to be stowed at FRP/section bases, or nearby, in secure stowages which are readily accessible in an emergency.

b. The large TIC must have its visor fitted before use.

c. Ni-cad rechargeable batteries are to be used for training purposes (duration approximately 45 minutes) and Duracell MN1500 batteries (duration approximately 90 minutes) are to be fitted before use in an emergency.

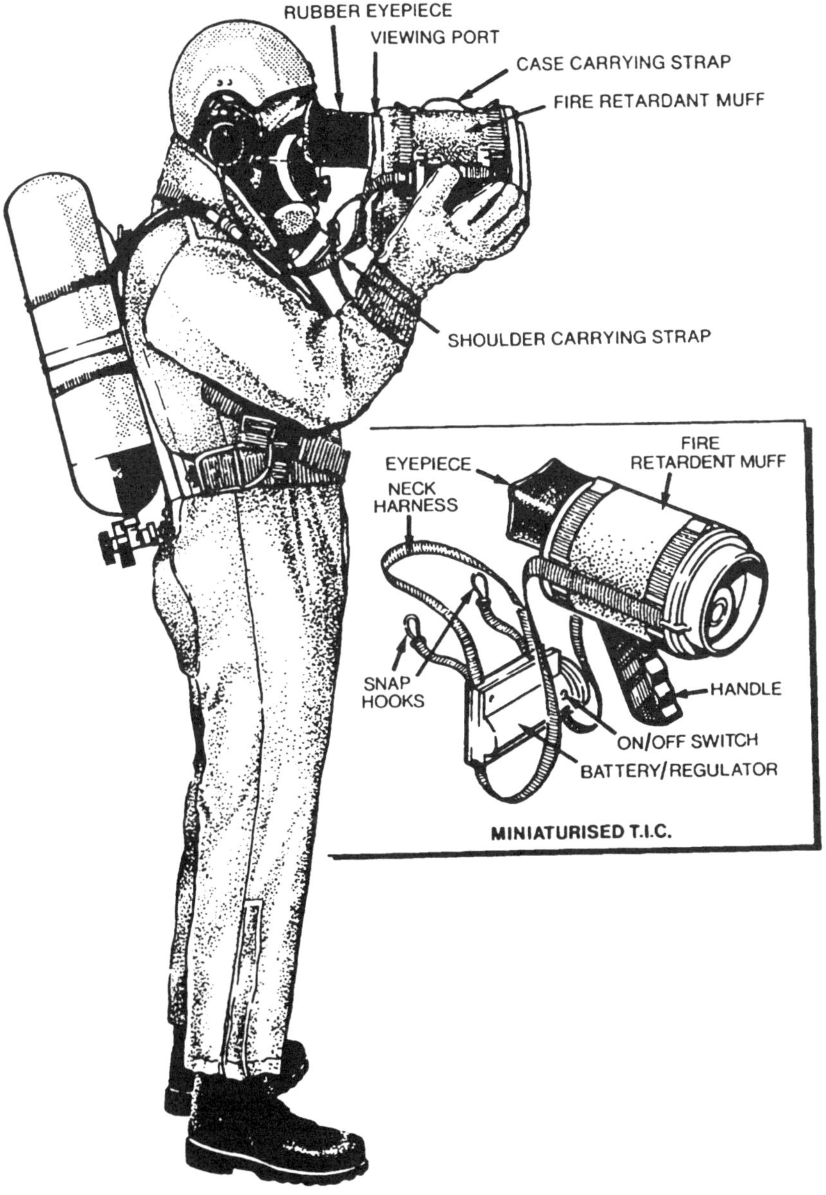

Fig 7-5. Thermal Imaging Camera

0707. Communications - General

Each FRP will be provided with ten sets (for 2 x 4 men firefighting teams, 1 for the FRP and 1 spare) of:

a. Portable radio transmitter/receivers (transceivers) comprising a radio body and battery section enclosed in a waterproof case and fitted with a flexible antenna. They are supplied with instructional manuals, and

b. Slim tank helmets (STH) (see Fig 7-6) fitted with an integral dual sided ear shells, a boom microphone, attachment sockets, and a press-to-speak (pressel) switch with a push in connector to the portable transceiver. There is an attachment for a DC caplamp (see Figs 7-1, 7-2 and 7-6).

These items are to be placed together in a secure locker which is readily accessible in an emergency. Where this is not possible the transceivers are to be stowed in the FRP lockers and the slim tank helmets stowed, in their carrying bags, nearby.

0708. Safety Equipment

a. A distress signal (bulb horn), an armband and BA control board are stowed with the BA in its locker. During firefighting operations the BA controller retains the armband from each set under his control and also operates the BA control board. **A maximum of 8 sets may be controlled by each controller**. The bulb horn, worn by each firefighter is clipped to the plastic D ring of the BA harness on the left hand side of the chest (see Fig 7-1).

b. Guide lines are stowed in BASCCA lockers adjacent to FRP and FP posts (including upper deck re-entry FPP). Personal lines are attached to each BASCCA. Both are to be used as directed by whoever is in charge of SAR teams during SAR operations.

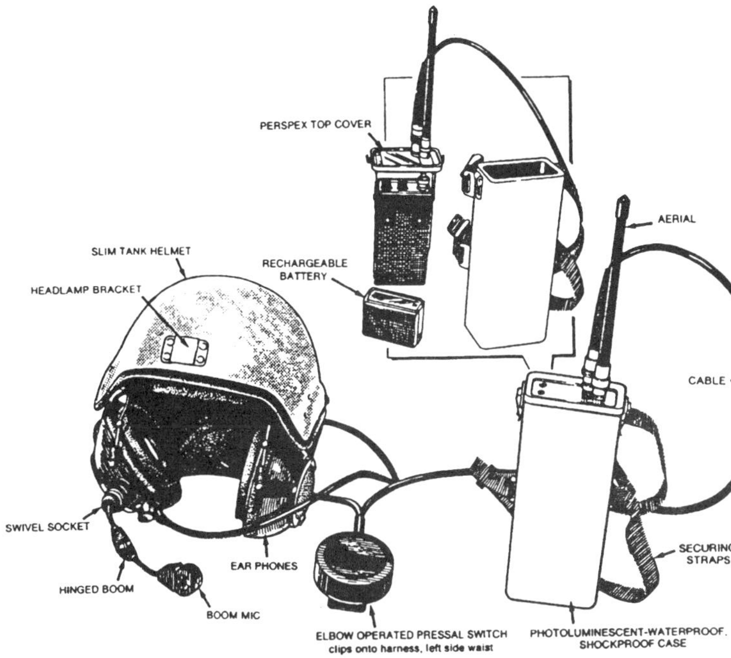

Fig 7-6. Slim Tank Helmet Kit

BREATHING APPARATUS, SELF CONTAINED, COMPRESSED AIR (DC&FF)

0709. General

There are 2 variants of BASCCA in service, the standard set for general use and the non magnetic set issued to HUNT and SANDOWN class. The non magnetic version has slightly less air capacity than the standard version and the BA Control Board (**Fig 7-7**) must be amended in accordance with the table at **para 0719.**

0710. Stowage

This equipment is stowed in bulkhead mounted lockers (see para 0705a) together with anti-flash hood and gloves. A tally detailing the functional and face seal checks is attached to the inside of the door. The door can be removed to prevent obstruction. The set must always be charged to 207 bar before stowage.

0711. Description

BASCCA comprises a facemask assembly, a frame and harness assembly, an air cylinder and valve assembly, a high pressure reducing valve, a warning whistle and a pressure gauge. The face mask assembly is fitted with an ori-nasal mask, an exhalation valve and a demand valve assemblies. The demand valve is designed to permit the use of the apparatus as a positive pressure set for normal operations, to ensure that any leaks in the face seal are outwards. The frame comprises a backplate with harness attachments and cylinder cradles. The cylinder has a free air capacity of 1400 litres in the standard set and 1210 in the non magnetic version. The maximum working pressure is 207 bar. When fully charged the standard set gives a nominal endurance to the whistle sounding of 27 minutes (25 for the non mag version), with a warning whistle set to sound when there is 7 minutes worth of air remaining. The pressure at which the whistle sounds is indicated by the red segment on the pressure gauge.

a. When BASCCA is worn the wearer must try to control his breathing so that it is rhythmic and steady. Breathing fast, or panting will greatly reduce the wear time.

b. Whenever possible it must be arranged that personnel with beards or small faces are not required to wear BASCCA(DC&FF) operationally as this reduces the endurance. In wartime all beards are to be shaved off.

c. Remember not to over tighten the facemask harness straps, and that all the air connections are designed to be hand tight.

0712. Extension Equipment

This comprises:

- A 2 metre extension hose.
- A facemask assembly and harness; and
- A 20 metre extension hose.

These items are stowed in a red painted locker which has a drop down door. The tally on the outside of the door reads EXTENSION EQUIPMENT FOR BASCCA (DC&FF), the tally on the inside gives the operating instructions.

OPERATING THE BASCCA (DC&FF)

0713. Donning/Functional Check

The operator, assisted by his controller, should don the apparatus as follows:

a. Remove the facemask from its stowage clip and place the neck strap over the head so allowing the facemask to hang on the chest. The neck strap must always be outside the operator's clothing and not in contact with the skin.

b. Release the apparatus securing bands and pass the right arm through the right shoulder strap; lift the apparatus clear of the locker and pass the left arm through the left shoulder strap.

c. Adjust the shoulder straps by pulling backwards towards the cylinder, then fasten the waist belt and chest strap. Do not over tighten the shoulder straps otherwise the waist belt may be too high.

d. Select negative pressure mode using the change over knob.

e. Open the cylinder valve fully. The warning whistle should sound briefly. If the whistle is not heard then shut the valve, select positive pressure mode and reduce pressure in the HP air hose to 43 bar. If the whistle still does not operate it is defective and this fact must be reported to the BA co-ordinator.

f. Select negative pressure mode and then open the cylinder valve. Should the air then leak from the demand valve the apparatus is faulty and must not be used. The BA co-ordinator is to be informed.

g. The BA controller removes and retains the control armband and records the set number and the operator's name on the control board.

0714. Face Seal Checks

When the operator is required to commence his task the following checks are to be carried out:

a. Don the facemask and, when it is fully adjusted for correct fitting, select the positive pressure mode and then commence to breathe from the apparatus. Ensure that the head harness is central at the top/back of the head by first adjusting the two lower straps at the same time and then the two upper straps.

b. Breathe deeply to ensure that the demand and exhalation valves are working correctly.

c. Observe the pressure gauge; close the cylinder valve and then open it half a turn, take and hold, a deep breath; then close the valve. Check that the pressure gauge reading does not drop by more than 5 bar in 10 seconds. Open the cylinder valve fully.

d. Check that the bulb horn works. It is only to be used for signalling an emergency.

Note. *The facemask sealing check is to be carried out whenever an operator dons the facemask for operational purposes.*

0715. Breathing Apparatus Control

When the BA's are used operationally a control number must be detailed. The BA controller must be well trained in the use of the apparatus and understand the important part he plays in the safety arrangements for the firefighters in the fire zone. He must always be in possession of a reliable timepiece and chinagraph pencils.

0716. The BA controller must remain outside the fire zone, near to the smoke boundary and adjacent to the rating in charge at the SOF. He must on no account be detailed for any other duties. He may act as the controller for up to eight men (two four men teams), if they are carry out the same task, but where possible the preferred number is four.

0717. The duties of a breathing apparatus controller are as follows:

a. Wear all the armbands of the BA's he is controlling.

b. Assist the operator(s) to don and adjust their BA and to carry out functional and face seal checks (see paras 0713 and 0714).

c. Mark up the control board (Form S3047) (see para 0718).

d. When its use is directed secure the tail of the guide line to convenient anchorage at the point of entry to the fire zone or incident.

e. Keep a sharp listening watch for any distress signals from the bulb horns.

f. Ensure that the BA controller of the relief firefighting team and the rating i/c of the Support Party are informed of the time the relief team should enter the fire zone in order to relieve the previous team.

g. Inform the rating in charge at the SOF when an operator has not returned at his **time due out**.

h. Tend the apparatus when it is used in conjunction with a 20 metre extension hose and facemask.

i. When controlling an operator, who is using the 20 metre extension hose, use an agreed signal to recall him.

j. Replace the control armbands over the pressure gauges when the operators report back from the fire zone/incident area.

0718. Completion of Control Board (Form S3047)

The BA controller carries out the following instructions after the firefighting team has been briefed by the officer/rating i/c of firefighting.

a. Complete columns A and B of the control board whilst firefighters are dressing (before the briefing, if possible).

b. When the facemasks are donned and the operators commence breathing, the remainder of the control board is to be completed (see Fig 7-7). All control board instructions must be carried out before the operators enter the fire zone/incident area.

c. When the operators return from the task, the BA controller must carry out the exit instructions as detailed on the control board.

d. When the operators return from the task they remove their facemasks with the change-over knob set to the negative pressure mode.

e. If re-entry to the fire zone/incident area is required, the operator and his BA controller must check the remaining cylinder pressure and complete the control board as instructed in a and b. He must also carry out the faceseal check.

0719. Control Board for Non Magnetic BASCCA

The Endurance table in the bottom RH corner of the S3047 must be amended as below when using the board for non magnetic BASCCA.

PRESSURE (bar)	TIME TO WHISTLE (min)
207	25
195	23
180	21
165	18
150	16
135	14
120	12

At 52 bar the whistle blows and 7 minutes breathing time remains.

Officer in Charge firefighters' permission required to start using BA if pressure is below 180 bar.

Control board for BASCCA (DCFF) 207 bar working pressure.

RN Form S3047 (Revised 10/85)

A			B	C	D	E	F	G
SET No.	OPERATOR'S NAME	TASK	CYLINDER PRESSURE (bar)	TIME TO WHISTLE (min)	TIME STARTED TO BREATHE	TIME RELIEF DUE IN	TIME RETURN COMMENCES	TIME DUE OUT
1	KEATING S.	FIREFIGHTER	207	27	1200	1200	1227	1234
27	ADAMS T	WATERWALL	200	25	1200	1218	1225	1232
13	HOBDAY	HOSE TENDER	195	25	1200	1218	1225	1232
6	SMALLEY	TEAM LEADER	185	23	1200	1216	1223	1230
8	SMITH K.	WATERWALL	207	27	1216	1236	1243	1250
9	McNAUGHTON	FIREFIGHTER	200	25	1216	1234	1241	1248
11	CURTIS J	TEAM LEADER	207	27	1216	1236	1243	1250
12	DENNIS J	HOSE TENDER	207	27	1216	1236	1243	1250

If operator does not emerge by time due out inform officer in charge firefighters

Entry Instructions

1. Remove control armband from set – to be worn by control number
2. Before operator enters complete A, B & D
3. Use Endurance Table to complete C
4. Add C to D to complete F
5. Subtract 7 minutes from F to complete E
6. Add 7 minutes to F to complete G
7. Check that the head harness is central at top/back of head
8. Ensure demand valve is in positive pressure mode
9. Carry out face seal check (pressure drop to be less than 5 bar in 10 seconds)

Exit Instructions

1. Remove armband and place over pressure gauge tube above clip
2. Operator to be debriefed by officer in charge firefighters then report to B.A. Co-ordinator

Endurance Table

PRESSURE (bar)	TIME TO WHISTLE (min)
207	27
195	25
180	23
165	20
150	18
135	15
120	13

At 43 bar the whistle blows and 7 minutes breathing time remains

Officer in charge firefighters' permission required to START using B.A. if pressure below ~~170~~ 180 bars

Note:
Times are approximate, actual duration will depend on individual and task

Fig 7-7. Breathing Apparatus Control Board - Form S3047

CHAPTER 8

FIREFIGHTING TACTICS UNIQUE TO SUBMARINES

Fig 8-1. Submarine Firefighting on Same Level - 3 Man Team with CFHR

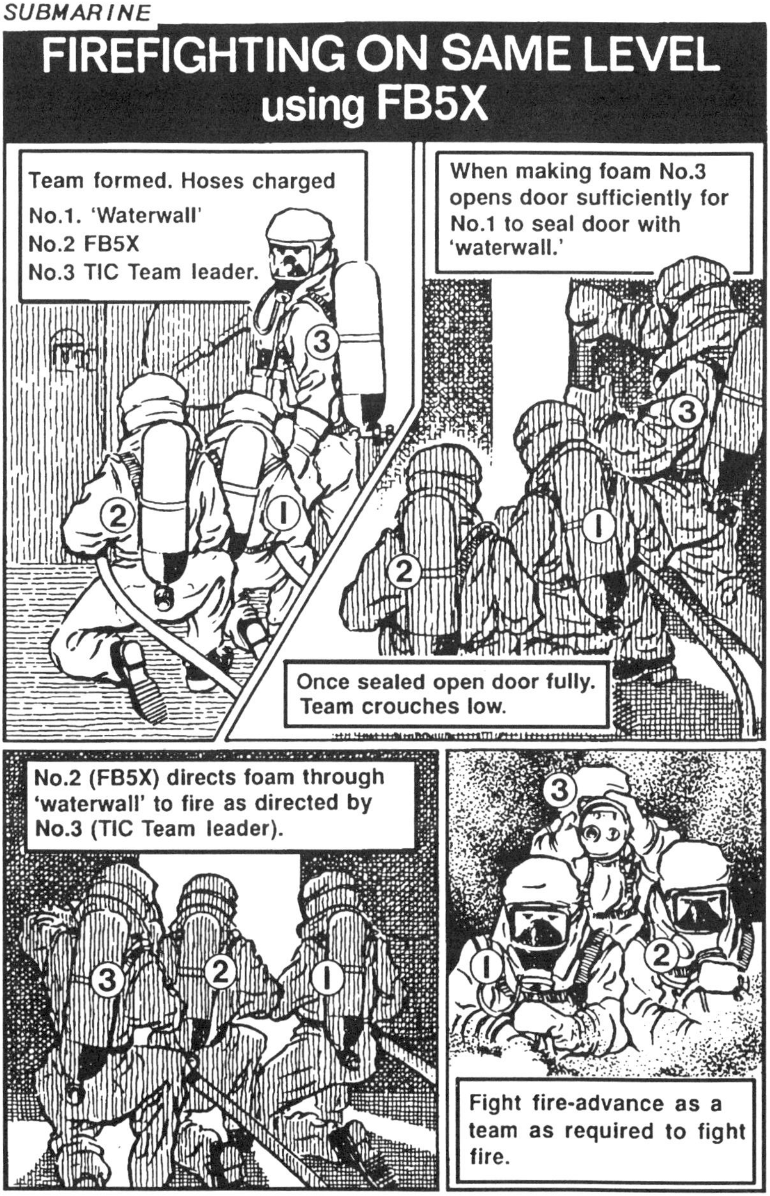

Fig 8-2. Submarine Firefighting on Same Level - 3 Man Team with FB5X

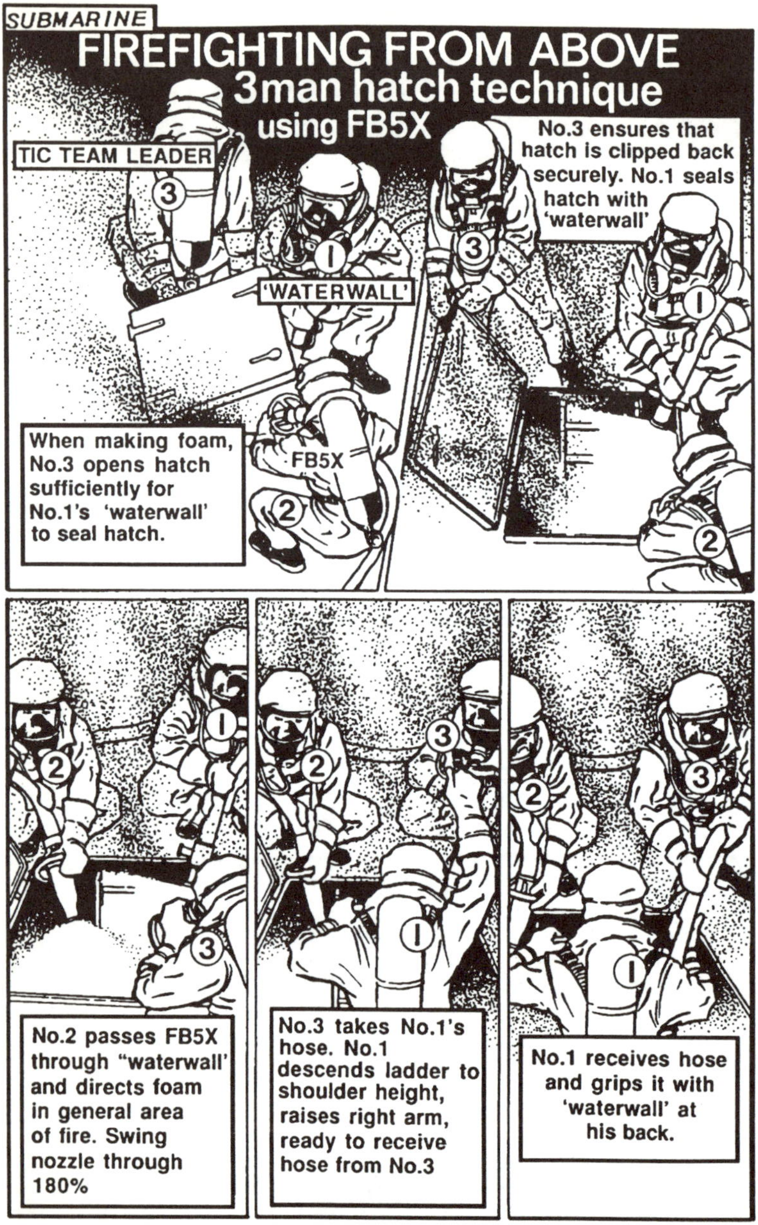

Fig 8-3a. Submarine Firefighting from Above
- 3 man team with 5BX

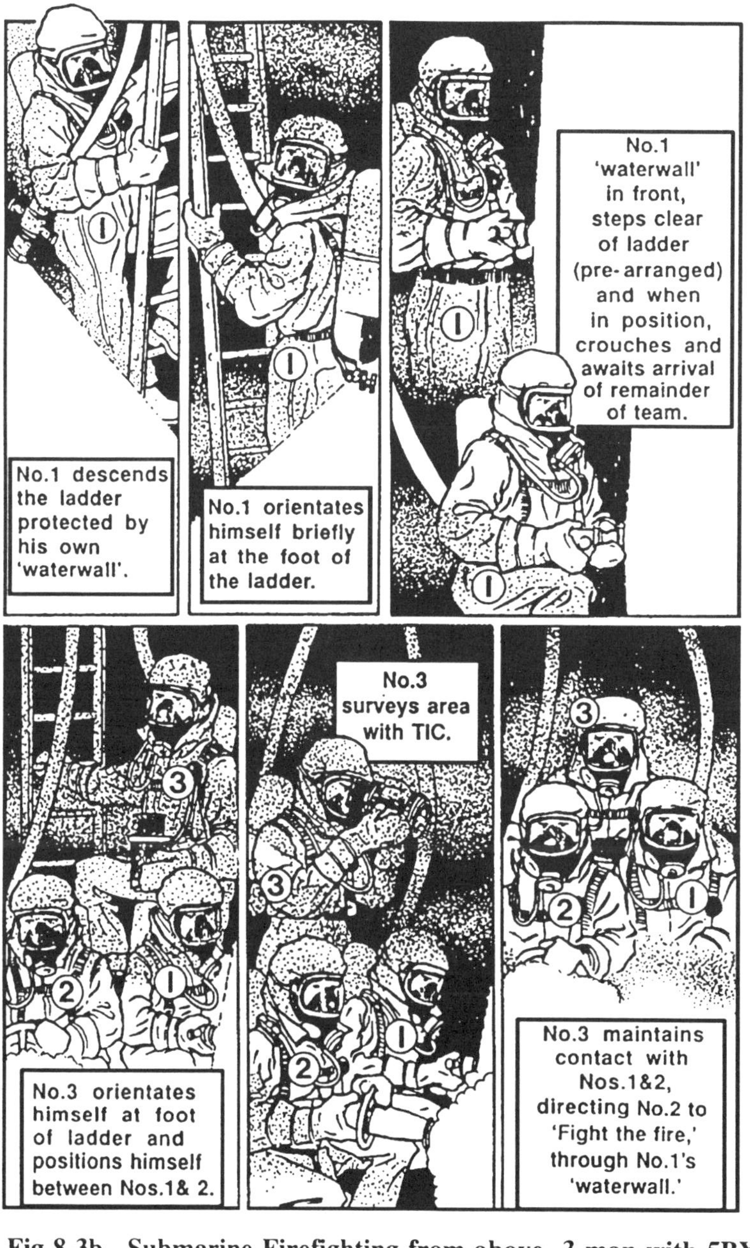

Fig 8-3b. Submarine Firefighting from above- 3 man with 5BX

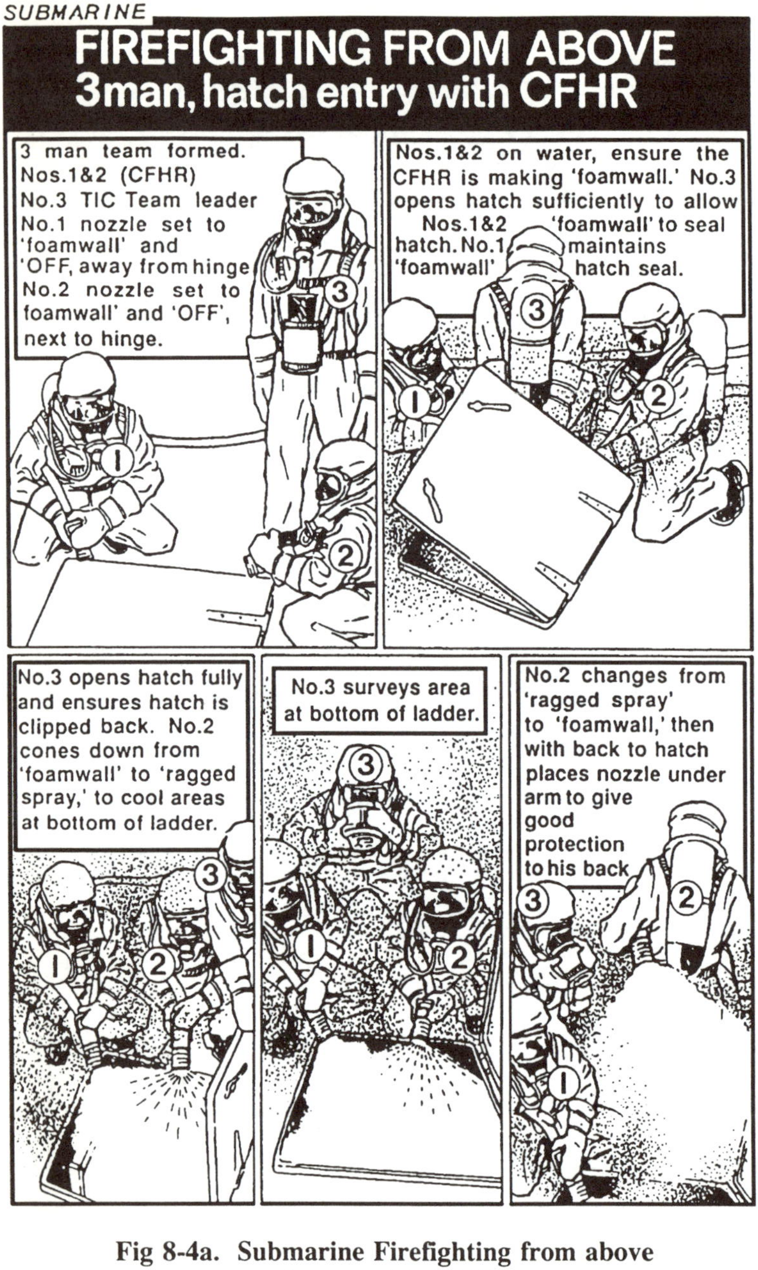

Fig 8-4a. Submarine Firefighting from above - 3 man with CFHR

Fig 8-4b. Submarine Firefighting from above
- 3 CFHR

Fig 8-5a. Submarine Firefighting from Below - 2 Man with CFHR

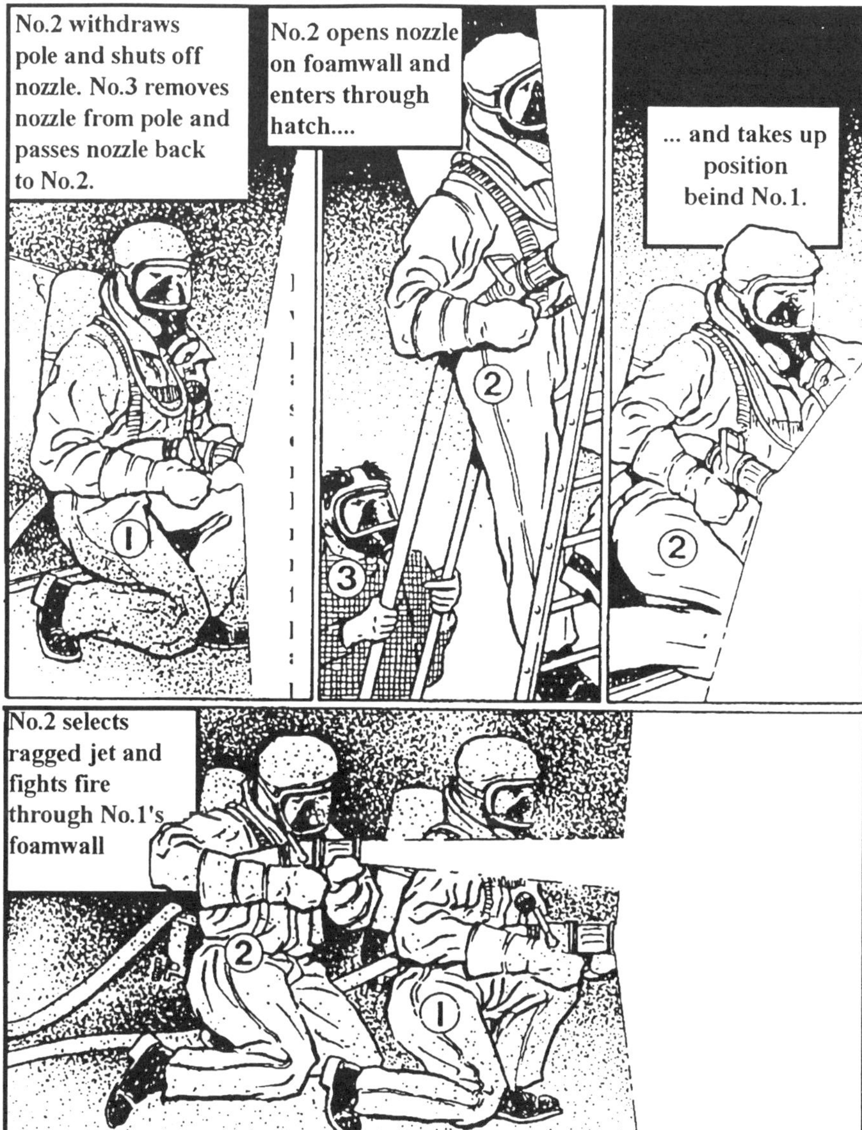

Fig 8-5b. Submarine firefighting from below - 2 man with CFHR

INDEX

Printed in the United Kingdom for HMSO
Dd. 296897 C108 12/94 9385